Compendium of Plant Genomes

Series Editor

Chittaranjan Kole, Raja Ramanna Fellow, Government of India, ICAR-National Research Center on Plant Biotechnology, Pusa, New Delhi, India

Whole-genome sequencing is at the cutting edge of life sciences in the new millennium. Since the first genome sequencing of the model plant Arabidopsis thaliana in 2000, whole genomes of about 100 plant species have been sequenced and genome sequences of several other plants are in the pipeline. Research publications on these genome initiatives are scattered on dedicated web sites and in journals with all too brief descriptions. The individual volumes elucidate the background history of the national and international genome initiatives; public and private partners involved; strategies and genomic resources and tools utilized; enumeration on the sequences and their assembly; repetitive sequences; gene annotation and genome duplication. In addition, synteny with other sequences, comparison of gene families and most importantly potential of the genome sequence information for gene pool characterization and genetic improvement of crop plants are described.

Interested in editing a volume on a crop or model plant? Please contact Prof. C. Kole, Series Editor, at ckoleorg@gmail.com

More information about this series at http://www.springer.com/series/11805

Chittaranjan Kole · Hideo Matsumura ·
Tusar Kanti Behera
Editors

The Bitter Gourd Genome

Springer

Editors
Chittaranjan Kole
Raja Ramanna Fellow
Government of India
ICAR-National Research
Center on Plant, Pusa
New Delhi, India

Hideo Matsumura
Gene Research Center
Shinshu University
Ueda, Japan

Tusar Kanti Behera
Division of Vegetable Science
ICAR-Indian Agricultural Research
Institute
New Delhi, India

ISSN 2199-4781   ISSN 2199-479X   (electronic)
Compendium of Plant Genomes
ISBN 978-3-030-15064-8   ISBN 978-3-030-15062-4   (eBook)
https://doi.org/10.1007/978-3-030-15062-4

This Springer imprint is published by the registered company Springer Nature Switzerland AG
The registered company address is: Gewerbestrasse 11, 6330 Cham, Switzerland

This book series is dedicated to my wife Phullara, and our children Sourav and Devleena
Chittaranjan Kole

Preface to the Series

Genome sequencing has emerged as the leading discipline in the plant sciences coinciding with the start of the new century. For much of the twentieth century, plant geneticists were only successful in delineating putative chromosomal location, function, and changes in genes indirectly through the use of a number of "markers" physically linked to them. These included visible or morphological, cytological, protein, and molecular or DNA markers. Among them, the first DNA marker, the RFLPs, introduced a revolutionary change in plant genetics and breeding in the mid-1980s, mainly because of their infinite number and thus potential to cover maximum chromosomal regions, phenotypic neutrality, absence of epistasis, and codominant nature. An array of other hybridization-based markers, PCR-based markers and combinations of both technologies, facilitated construction of genetic linkage maps, genetic mapping of genes controlling simply inherited traits, and even gene clusters controlling polygenic traits (QTLs) in a large number of model and crop plants. During this period, a number of new mapping populations beyond F_2 were utilized, and a number of computer programs were developed for map construction, mapping of genes, and for mapping of polygenic clusters or QTLs. Molecular markers were also used in the studies of evolution and phylogenetic relationship, genetic diversity, DNA fingerprinting, and map-based cloning. Markers tightly linked to the genes were used in crop improvement employing the so-called marker-assisted selection. These strategies of molecular genetic mapping and molecular breeding made a spectacular impact during the last one and a half decades of the twentieth century. But still they remained "indirect" approaches for elucidation and utilization of plant genomes since much of the chromosomes remained unknown, and the complete chemical depiction of them was yet to be unraveled.

Physical mapping of genomes was the obvious consequence that facilitated the development of the "genomic resources" including BAC and YAC libraries to develop physical maps in some plant genomes. Subsequently, integrated genetic–physical maps were also developed in many plants. This led to the concept of structural genomics. Later on, the emphasis was laid on EST and transcriptome analysis to decipher the function of the active gene sequences leading to another concept defined as functional genomics. The advent of techniques of bacteriophage gene and DNA sequencing in the 1970s was extended to facilitate sequencing of these genomic resources in the last decade of the twentieth century.

As expected, the sequencing of chromosomal regions would have led to too much data to store, characterize, and utilize with the-then available computer software could handle. But the development of information technology made the life of biologists easier by leading to a swift and sweet marriage of biology and informatics, and a new subject was born—bioinformatics.

Thus, the evolution of the concepts, strategies, and tools of sequencing and bioinformatics reinforced the subject of genomics—structural and functional. Today, genome sequencing has traveled much beyond biology and involves biophysics, biochemistry, and bioinformatics!

Thanks to the efforts of both public and private agencies, genome sequencing strategies are evolving very fast, leading to cheaper, quicker, and automated techniques right from clone-by-clone and whole-genome shotgun approaches to a succession of the second-generation sequencing methods. The development of software of different generations facilitated this genome sequencing. At the same time, newer concepts and strategies were emerging to handle sequencing of the complex genomes, particularly the polyploids.

It became a reality to chemically—and so directly—define plant genomes, popularly called whole-genome sequencing or simply genome sequencing.

The history of plant genome sequencing will always cite the sequencing of the genome of the model plant Arabidopsis thaliana in 2000 that was followed by sequencing the genome of the crop and model plant rice in 2002. Since then, the number of sequenced genomes of higher plants has been increasing exponentially, mainly due to the development of cheaper and quicker genomic techniques and, most importantly, the development of collaborative platforms such as national and international consortia involving partners from public and/or private agencies.

As I write this preface for the first volume of the new series "Compendium of Plant Genomes," a net search tells me that complete or nearly complete whole-genome sequencing of 45 crop plants, eight crop and model plants, eight model plants, 15 crop progenitors and relatives, and 3 basal plants is accomplished, the majority of which are in the public domain. This means that we nowadays know many of our model and crop plants chemically, i.e., directly, and we may depict them and utilize them precisely better than ever. Genome sequencing has covered all groups of crop plants. Hence, information on the precise depiction of plant genomes and the scope of their utilization are growing rapidly every day. However, the information is scattered in research articles and review papers in journals and dedicated Web pages of the consortia and databases. There is no compilation of plant genomes and the opportunity of using the information in sequence-assisted breeding or further genomic studies. This is the underlying rationale for starting this book series, with each volume dedicated to a particular plant.

Plant genome science has emerged as an important subject in academia, and the present compendium of plant genomes will be highly useful both to students and teaching faculties. Most importantly, research scientists involved in genomics research will have access to systematic deliberations on the plant genomes of their interest. Elucidation of plant genomes is of interest not only for the geneticists and breeders, but also for practitioners of an array of plant science disciplines, such as taxonomy, evolution, cytology, physiology, pathology, entomology, nematology, crop production, biochemistry, and obviously bioinformatics. It must be mentioned that information regarding each plant genome is ever-growing. The contents of the volumes of this compendium are, therefore, focusing on the basic aspects of the genomes and their utility. They include information on the academic and/or economic importance of the plants, description of their genomes from a molecular genetic and cytogenetic point of view, and the genomic resources developed. Detailed deliberations focus on the background history of the national and international genome initiatives, public and private partners involved, strategies and genomic resources and tools utilized, enumeration on the sequences and their assembly, repetitive sequences, gene annotation, and genome duplication. In addition, synteny with other sequences, comparison of gene families, and, most importantly, the potential of the genome sequence information for gene pool characterization through genotyping by sequencing (GBS) and genetic improvement of crop plants have been described. As expected, there is a lot of variation of these topics in the volumes based on the information available on the crop, model, or reference plants.

I must confess that as the series editor, it has been a daunting task for me to work on such a huge and broad knowledge base that spans so many diverse plant species. However, pioneering scientists with lifetime experience and expertise on the particular crops did excellent jobs editing the respective volumes. I myself have been a small science worker on plant genomes since the mid-1980s and that provided me the opportunity to personally know several stalwarts of plant genomics all over the globe. Most, if not all, of the volume editors are my longtime friends and colleagues. It has been highly comfortable and enriching for me to work with them on this book series. To be honest, while working on this series, I have been and will remain a student first, a science worker second, and a series editor last. And I must express my gratitude to the volume editors and the chapter authors for providing me the opportunity to work with them on this compendium.

I also wish to mention here my thanks and gratitude to the Springer staff particularly, Dr. Christina Eckey and Dr. Jutta Lindenborn for the earlier set of volumes and presently Ing. Zuzana Bernhart for all their timely help and support.

I always had to set aside additional hours to edit books beside my professional and personal commitments—hours I could and should have given to my wife, Phullara, and our kids, Sourav and Devleena. I must mention that they not only allowed me the freedom to take away those hours from them but also offered their support in the editing job itself. I am really not sure whether my dedication of this compendium to them will suffice to do justice to their sacrifices for the interest of science and the science community.

Kalyani, India   Chittaranjan Kole

Preface to the Volume

The incidence rate of several deadly diseases, specifically cancer and diabetes, is highly alarming. According to the report of the International Agency for Research on Cancer of World Health Organization, the global cancer burden is estimated to have risen to 18.1 million new cases and 9.6 million deaths in 2018. Worldwide, the total number of people who are alive within 5 years of a cancer diagnosis, called the 5-year prevalence, is estimated to be 43.8 million. According to the projections by the International Diabetes Federation, in 2017, approximately 425 million adults were living with diabetes; by 2045, this will rise to 629 million.

Utilization of medicinal plants and nutraceutical crops is the potential options for alternative and complimentary medicines to mitigate these problems. According to the Zion Market Research, the global herbal supplement market is expected to reach approximately USD 86.74 billion by 2022, growing at a CAGR of around 6.8% between 2017 and 2022. Presently, plant-based drugs contribute 50% to linical drugs. This commercial importance coupled with severe prevalence of the deadly diseases underscores the need for the generation of genetic, genomics, and breeding resources in medicinal plants and functional food crops.

Bitter gourd, also known as African cucumber, ampalaya, balsam pear, balsam apple, bitter apple, and bitter cucumber, is grown traditionally in the tropical and subtropical areas in Asia, South America (Amazon region), East Africa, and the Caribbean as a food vegetable and medicine. This plant contains over 60 phytomedicines potent against 30 diseases. Medicinal properties of bitter melon including antidiabetic, antiviral, antitumor, antileukemic, antibacterial, antihelmintic, antimutagenic, antimycobacterial, antioxidant, antiulcer, anti-inflammatory, hypocholesterolemic, hypotriglyceridemic, hypotensive, immunostimulant, and insecticidal properties have been well documented in research. All parts of this plant, mainly the fruits and the seeds, contain cucurbitacin-B, lycopene, and β-carotene, which are known to have anticancer actions. They also contain charantin and plant insulin that have been clinically demonstrated to have hypoglycemic and anti-hyperglycemic activities and established beneficial effects on diabetes, particularly of type-2. Antioxidant properties of bitter gourd and its traditional use in several countries worldwide including India have been well documented and reviewed in the literature.

Despite immense importance of bitter gourd as a medicinally important vegetable crop and its commercial value, it remained as an "underutilized" or "orphan" crop. Some serious efforts have been started only recently to scientifically demonstrate the medicinal properties of the bioactive compounds in this plant and their mode of actions, and utilization of molecular markers in the evaluation of genetic diversity and elucidation of the genetics of its agro-economic characters. Sequencing of the whole bitter melon genome has been done only in the recent past. Information generated from these molecular genetic and genomic studies will now facilitate breeding of elite varieties with higher fruit yield and content of phytomedicines.

Obviously, there is no book available with compiled information of the botanical descriptions, medicinal properties, genetic and genomic studies, and breeding in this important crop plant. We tried to perform this task by presenting 12 chapters in our book entitled The Bitter Gourd Genome contributed by us and many reputed scientists. Chapter 1 briefs the economic importance of bitter melon as a vegetable crop and also as a medicinal plant and provides glimpses on the works done so far on the areas of genetics, breeding, and genomics. Chapter 2 presents comprehensive information on the botany of the crop under the sections of origin and distribution, taxonomy, morphology, floral biology and mode of reproduction, sex phenology, anatomy, ecology, and economic botany. Chapter 3 deliberates on a variety of bioactive compounds present in bitter gourd including alkaloids, polypeptides, vitamins, and minerals and draws a link of the bioactive compounds to its pharmacological effects like antidiabetic, anticancer, antiviral, anti-inflammatory, analgesic, hypolipidemic, and hypocholesterolemic effects, and provides an insight to understand the mechanism of action. Information on the three gene pools and their potential as a resource of genes useful in breeding has been narrated in Chap. 4. In addition, genetic diversity with regard to morphological characters and content of various phytomedicines has also been delineated. Chapter 5 includes a description of karyotype and chromosomal configurations in this crop and also advanced results from cytomolecular investigations. Cucurbits constitute a unique family including crops with the phenomenon of sex determination. Chapter 6 illuminates the genetic and genomic studies of sex determination in bitter gourd and highlights their potential in elucidating evolution of monoecy and dioecy and more importantly the use of gynoecious lines on crossbreeding. Almost no serious research has been conducted on biotechnology in this crop. However, available information on *in vitro* culture and nanotechnology has been reviewed in Chap. 7. Elaborate discussions on classical genetics and traditional breeding have been covered in Chap. 8 under relevant sections including genetics of agro-economically important qualitative and quantitative traits and details on strategies and tools of conventional breeding. Chapter 9 focuses on various molecular markers and their use in construction of genetic linkage map in bitter gourd. It also deliberates on mapping of many important simply inherited and polygenic traits (QTLs) on these maps. Chapter 10 enumerates on the sequencing of the bitter gourd genome, annotation, and comparison of this genome and its annotation with those in other Cucurbitaceae species that establishes phylogenetical distance of bitter gourd from other known cucurbit crops and, also the unique properties conferred by the encoding genes, specifically the RIP genes underlying the antitumor or antiviral activities. Since genomics studies have been initiated in bitter gourd much later than many crop plants, only limited works have been done on metabolomics despite of its importance to substantiate the molecular mechanisms of the secondary metabolites such as flavonoids, phenolics, sterols, and terpenoids in this plant in conferring medicinal activities. Chapter 11 briefs the studies on the role of such metabolies, particularly terpenois, on imparting bitterness, medicinal properties, and host responses to pathogens and predators. Finally, a future road has been depicted in Chap. 12 for developing high-density molecular maps, genome sequence with more coverage utilizing advanced sequencing strategies and bioinformatics tools, and ultimately using this information in precise breeding in this crop.

New Delhi, India   Chittaranjan Kole
Ueda, Japan   Hideo Matsumura
New Delhi, India   Tusar Kanti Behera

Contents

Contributors

Vidhu Aeri School of Pharmaceutical Education and Research, New Delhi, India

A. C. Asna Department of Plant Breeding and Genetics, Kerala Agricultural University, Thrissur, Kerala, India

Tusar Kanti Behera Division of Vegetable Sciences, ICAR-Indian Agricultural Research Institute, Pusa, New Delhi, India

Sutapa Datta ICAR-National Institute for Plant Biotechnology, Pusa, New Delhi, India

Shyam Sundar Dey Division of Vegetable Sciences, ICAR-Indian Agricultural Research Institute, Pusa, New Delhi, India

Takeshi Furuhashi Anicom Specialty Medical Institute Inc, Tokyo, Japan

K. K. Gautam ICAR-Indian Institute of Vegetable Research, Varanasi, India

Gograj Singh Jat Division of Vegetable Science, ICAR-Indian Agricultural Research Institute, New Delhi, India

Jiji Joseph Department of Plant Breeding and Genetics, Kerala Agricultural University, Thrissur, Kerala, India

K. Joseph John ICAR-NBPGR Regional Station, Thrissur, Kerala, India

Khalid Mahmood Khawar Department of Crop Science, Faculty of Agriculture, Ankara University, Ankara, Turkey

Chittaranjan Kole ICAR-National Institute for Plant Biotechnology, Pusa, New Delhi, India

Ricardo A. Lombello Center for Natural and Human Sciences, Federal University of ABC, São Bernardo do Campo, SP, Brazil

Hideo Matsumura Gene Research Center, Shinshu University, Ueda, Nagano, Japan

Sudhakar Pandey ICAR-Indian Institute of Vegetable Research, Varanasi, India

Mamta Pathak Department of Vegetable Science, Punjab Agricultural University, Ludhiana, India

Richa Raj School of Pharmaceutical Education and Research, New Delhi, India

Sevil Saglam Yilmaz Department of Agricultural Biotechnology, Faculty of Agriculture, Kirsehir Ahi Evran University, Kirsehir, Turkey

Naoya Urasaki Okinawa Agricultural Research Center, Itoman, Okinawa, Japan

Abbreviations

½ ×MS medium	Half strength Murashige and Skoog's medium
2,4-D	2, 4 Dichlorophenoxyacetic acid

3 Medicinal

6-BA	6-Benzyladenine
6-BAP	6-Benzylaminopurine
a.a.	Amino acid
ABA	Abscisic acid
ABG-6 medium	½ × MS medium containing 0.5 mg/l BAP
ACC	Aminocyclopropane-1-carboxylic acid
ACS	Aminocyclopropane-1-carboxylic acid synthase
AFLP	Amplified fragment length polymorphism
AKT	Protein kinase
ALT	Alanine aminotransferase
AMPK	Adenosine-5-monophosphate kinase
Apo-A-1	Apolipoprotien A-1
Apo-B	Apolipoprotien B
AP-PCR	Arbitrary primed polymerase chain reaction
AST	Aspartate aminotransferase
ATL	Adult T-cell leukemia
BDMV	Bitter gourd distortion mosaic virus
BLAST	Basic Local Alignment Search Tool
BUSCO	Benchmarking universal single-copy orthologs
Bwa	Burrows–Wheeler Aligner
Cas9	CRISPR-associated protein 9
CCR-B	Cuicurbitacin-B
cDNA	Complementary DNA
CdS	Cadmium sulfide
CDSs	Coding sequences
CHR	Charantin
CMA	Chromomycin A_3
CRISPR	Clustered regularly interspaced short palindromic repeats
CVD	Cardiovascular disease
DAF	DNA amplification fingerprinting
DAMPs	Damage-associated molecular patterns
DAPI	4′,6-Diamidino-2-phenylindole
DaRT	Diversity array technology

DPPH	2,2-Diphenyl-1-picrylhydrazyl
DPPH	1,1-Diphenyl-2-picrylhydrazyl
FIASCO	Functional image analysis software-computational olio
FISH	Fluorescence *in situ* hybridization
FRAP	Ferric reducing ability of plasma
GAE	Gallic acid equivalents
GBS	Genotyping by sequencing
GCA	General combining ability
GC-MS	Gas chromatography–mass spectrometry
GCV	Genotypic coefficient of variation
GD	Genetic distance
GLUT-4	Glucose transport type 4
GP 1	Primary gene pool
GP 2	Secondary gene pool
GP 3	Tertiary gene pool
GSK-3	Glycogen synthase kinase-3
GUS test	β-Glucuronidase test
HDL	High-density lipoprotein
Hep G2	Hepatocellular cancer cell lines
HER 2	Human epidermal growth factor receptor
HIV	Human immunodeficiency virus
HL 60	Human leukemia cells
HR	Hypersensitive response
IBA	Indole 3 butyric acid
IC50	Half maximal inhibitory concentration
IL-1b	Interleukin- 1 beta
IL-6	Interleukin- 6
indel	Insertion and deletion
ISSR	Inter-simple sequence repeat
IU	International unit
JA	Jasmonic acid
LDL	Low-density lipoprotein
LG	Linkage group
LOD	Logarithm of the odds
LPS	Lipopolysaccharide
MAMPs	Microbe-associated molecular patterns
MAP	Mitogen-activated protein
MAP 30	Momordica antiviral protein 30kD
MAS	Marker-assisted selection
matK	Maturase K
MC	*Momordica charantia*
MI	Marker index
MMP-2	Matrix metalloproteinases-2
MMP-9	Matrix metalloproteinases-9
MS medium	Murashige and Skoog medium
NAA	α Naphthalene acetic acid
NCBI	National Center for Biotechnology Information

NF-κB	Nuclear factor kappa-light-chain-enhancer of activated B cells
NGS	Next-generation sequencing
NMR	Nuclear magnetic resonance
nptII	Neomycin phosphotransferase II
ORFs	Open reading frames
PAMPs	Pathogen-associated molecular patterns
PAs	Polyamines
PCD	Programmed cell death
PCR	Polymerase chain reaction
PCV	Phenotypic coefficient of variation
PIC	Polymorphic information content
PPR	Pentatricopeptide repeat
PTEN	Phosphatase and tensin homolog
QTL	Quantitative trait locus
QTLs	Quantitative trait loci
QTL-seq	Quantitative trait locus sequencing
RAD	Restriction-associated DNA
RAD-seq	Restriction-associated DNA sequencing
RAD-tags	Restriction-associated DNA tags
RAPD	Random amplified polymorphic DNA
RBG7 medium	½ × MS medium containing 1 mg/l IBA
RFLP	Restriction fragment length polymorphism
RIP	Ribosome inactivating protein
RNA-seq	RNA sequencing
RP	Revolving power
SAR	Systemic acquired resistance
SCA	Specific combining ability
SCAR	Sequence characterized amplified regions
SCT	Seed coat tissue
SNP	Single nucleotide polymorphism
SNVs	Single nucleotide variants
SOD	Superoxide dismutase
Spd	Spermidine
Spm	Spermine
SRAP	Sequence-related amplified polymorphism
SSR	Simple sequence repeat
STS	Sequenced tagged site
t-BHP	Tert-butyl hydroperoxide
TCS	Trichosanthin
TDZ	1-Phenyl-3-(1,2,3-thiadiazol-5-yl)urea or thidiazuron
TLR	Toll-like receptor
TNF-α	Tumor necrosis factor alpha
TPS	Terpene synthase
VLDL	Very low-density lipoprotein
Zn-finger	Zinc finger
α-MMC	α-Momorcharin

Glimpse on Genomics and Breeding in Bitter Gourd: A Crop of the Future for Food, Nutrition and Health Security

Tusar Kanti Behera, Hideo Matsumura
and Chittaranjan Kole

Abstract

Bitter gourd, *Momordica charantia* L., family Cucurbitaceae, plausibly originated in eastern Asia, is traditionally cultivated as a vegetable and medicinal crop in tropical and subtropical areas in Asia, South America, East Africa, and the Caribbean. It has a simple genome with $2n = 22$ chromosomes having a genome size of around 339 Mb. All parts of this plant, mainly the fruits and the seeds, contain more than 60 phytomedicines active against more than 30 diseases including cancer and diabetes. Single plant selection, mass selection, pedigree selection, and bulk population improvement are common methods are used widely in the bitter gourd improvement program. Recent discovery of gynoecious lines and their genetics will facilitate hybrid breeding. Ample genetic diversity has been found to exist in this crop as assessed by the use of molecular markers. Association mapping led to the detection of molecular markers linked to some fruit traits and content of a couple of phytomedicines. A few molecular genetic maps have been constructed and a number of agroeconomically important qualitative and quantitative fruit traits have been mapped. Recently, a draft genome sequence has also been reported and a few studies on genotyping by sequencing and RAD sequencing have been accomplished.

The vegetable bitter gourd, *Momordica charantia* L., belongs to the family of Cucurbitaceae. It is known by various names in different parts of the world and most commonly called bitter gourd, balsam pear, bitter melon, bitter cucumber, and African cucumber (Heiser 1979). Besides its principal use as a vegetable crop in south, southeast, and east Asia, sometimes it is also cultivated as an ornamental and is very popular as folk medicine (Heiser 1979). The fruits are cooked in varied ways in different regions and they are mixed with a wide range of vegetable crops. Bitter gourd fruits are also used as stuffed, stir-fried vegetables and added in small quantities to beans and soups to provide a slightly bitter flavor. Sometimes, soaking the fruits in salt water is also practiced to reduce the bitterness of the fruits before cooking. The fruits are also processed as dehydrated sliced chips, pickled, or canned besides their use in culinary purposes. Besides fruits, flowers and young shoots are also used as flavoring agents in various Asian dishes.

T. K. Behera
Division of Vegetable Science, ICAR-Indian
Agricultural Research Institute, Pusa, New Delhi
110012, India

H. Matsumura
Gene Research Center, Shinshu University, Ueda,
Nagano, Japan

C. Kole (✉)
ICAR-National Institute for Plant Biotechnology,
Pusa, New Delhi 110012, India
e-mail: ckoleorg@gmail.com

© Springer Nature Switzerland AG 2020
C. Kole et al. (eds.), *The Bitter Gourd Genome*, Compendium of Plant Genomes,
https://doi.org/10.1007/978-3-030-15062-4_1

Use of the tender *Momordica* shoots and leaves as leafy vegetables and cooking is also reported. In some regions, leaf and fruit extracts are used in the preparation of tea (Tindall 1983; Reyes et al. 1994). Unlike other cucurbitaceous vegetables, the bitter fruit flavor of *M. charantia* is considered desirable for consumption, and thus, bitter flavor has been selected during domestication (Marr et al. 2004).

Bitter gourd is a diploid species with somatic chromosome number of $2n = 22$ and its approximate genome size is around 339 Mb as per its draft genome sequence (Urasaki et al. 2017). It is an important crop of the tropical and subtropical areas in Asia, South America, East Africa, and the Caribbean, which is used as a vegetable and medicine. It is grown widely in India, Sri Lanka, Philippines, Thailand, Malaysia, China, Japan, Australia, Tropical Africa, South America, and the Caribbean. In India, Tamil Nadu, Uttar Pradesh, Maharashtra, Kerala, and Karnataka are the important bitter gourd producing states. The center of origin of bitter gourd is most likely in eastern Asia, possibly eastern India, or southern China (Walters and Decker-Walters 1988; Miniraj et al. 1993). Further, it was domesticated in Asia, possibly in eastern India or southern China. The uncarbonized seed coat fragments have been tentatively identified from Spirit Cave in northern Thailand. However, there have been no archaeological reports of bitter gourd remains in China (Marr et al. 2004). Moreover, a comprehensive compilation of plant remains from 124 Indian archaeological sites does not include bitter gourd (Kajale 1991). Wild or small-fruited cultivated forms, however, are mentioned in Ayurvedic texts written in Indian Sanskrit from 2000 to 200 BCE by members of the Indo-Aryan culture (Decker-Walters 1999), indicating an early cultivation of bitter gourd in India. The lack of a unique set of Indo-Aryan words indicates that the Aryans did not know bitter gourd before entering India (Walters and Decker-Walters 1988). The earliest written reference to bitter gourd in China was made in 1370 CE (Yang and Walters 1992). Both the domesticated and putative wild bitter gourd progenitors of bitter gourd are listed in

floras of India, tropical Africa, and Asia as well as the New World tropics, where it first arrived in Brazil via the slave trade from Africa and then spread into Central America (Marr et al. 2004). Based on both historical literature (Walters and Decker-Walters 1988; Chakravarty 1990; Miniraj et al. 1993) and recent molecular analyses employing random amplified polymorphic DNA (RAPD; Dey et al. 2006), inter-simple sequence repeats (ISSR; Singh et al. 2007) and amplified fragment length polymorphisms (AFLP; Gaikwad et al. 2008) molecular analyses, eastern India may be considered as a probable primary center of diversity of bitter gourd, where a wild feral form *M. charantia* var. *muricata* (Chakravarty 1990) currently exists. *M. charantia* and *M. balsamina* are monoecious annuals while the tuberous perennials, *M. dioica*, *M. subangulata* ssp. *renigera*, *M. cochinchinensis*, *M. sahyadrica*, *M. foetida,* and *M. rostrata* are dioecious. However, hermaphrodite flowers have been observed in *M. dioica* (Jha and Roy 1989), *M. charantia,* and *M. subangulata* ssp. *renigera*. Gynoecious lines originating in India were identified by Behera et al. (2006; lines DBGy-201 and DBGy-202) and Ram et al. (2002; line Gy263B) for use in hybrid development programs.

Medicinal properties of bitter gourd including antidiabetic, antiviral, antitumor, antileukemic, antibacterial, antihelmintic, antimutagenic, antimycobacterial, antioxidant, antiulcer, anti-inflammatory, hypocholesterolemic, hypotriglyceridemic, hypotensive, immunostimulant, and insecticidal properties have been well documented (Raman and Lau 1996; Basch et al. 2003; Behera et al. 2010). All parts of this plant, mainly the fruits and the seeds, contain cucurbitacin-B, lycopene, and β-carotene, which are known to have anticancer actions (Grover and Yadav 2004; Kole et al. 2013). All these parts also contain charantin and plant insulin (syn. polypeptide-p) (Kole et al. 2013) that have been clinically demonstrated to have hypoglycemic and antihyperglycemic activities (Grover et al. 2002) and established beneficial effects on diabetes, particularly of type-2. Antioxidant properties of bitter gourd and its traditional use in several countries worldwide including India have been well reviewed in the literature (Sathishsekar and Subramanian 2005; Semiz and Sen 2007; Behera et al. 2010).

Momordica charatia is predominantly monoecious in nature where staminate and pistillate flowers are borne on separate nodes. However, Roy (1973) has reported the appearance of various intermediate sex forms like andromonoecious, gynoecious, and trimonoecious in colchicine treated plants of bitter gourd without changing their ploidy level. In the last decade, gynoecism has been reported in bitter gourd in India, Japan, and China (Ram et al. 2002; Behera et al. 2006; Iwamoto and Ishida 2005). Gynoecism in bitter gourd is under the control of a single, recessive gene (*gy-1*) (Behera et al. 2009) and has been used widely in development of predominantly gynoecious lines and gynoecious-based F_1 hybrids. Recently, two flanking markers, TP_54865 and TP_54890 on LG 12 at a distance of 3.04 cM to TP_54890, have been identified for the gynoecious (*gy-1*) locus (Gangadhara Rao et al. 2018). Sex expression is affected by environmental conditions under which bitter gourd seedlings grow (Wang and Zeng 1997). Short-day cultivars when growing under short photoperiods exhibit rapid development and comparatively high gynoecy. To encourage a high frequency of pistillate flowers, such short-day treatments should begin at seedling emergence and proceed to sixth-leaf stage (~ 20 days post-emergence under growing optimal conditions). While low temperature enhances short-day effects, relatively high temperatures typically delay reproductive growth, weakening short-day responses. Likewise, pistillate flower production under short days is increased by low temperatures (e.g., 20 °C) and nighttime chilling (e.g., 25 °C day/15 °C night) (Yonemori and Fujieda 1985). Consequently, optimal conditions for bitter gourd seedling growth are short days and low temperatures (Wang and Zeng 1997).

Several breeding methods are employed in bitter gourd to accomplish varied breeding objectives. Single plant selection, mass selection, pedigree selection, and bulk population improvement are common methods that are used widely in bitter gourd improvement program (Sirohi 1997). Pedigree selection typically is used after crossing two parents for the development of inbred lines with high, early yield borne on a unique plant habit, and/or with high-quality fruit (i.e., processing quality, high vitamin C and A, and disease resistance). The first genetic map and positions of major fruit trait loci of bitter gourd were worked out by Kole et al. (2012). Thereafter, an extensive genetic linkage map was constructed via the study of $F_{2:3}$ progenies derived from two cultivated inbred lines (Wang and Xiang 2013). Matsumura et al. (2014) identified a single nucleotide polymorphism (SNP) marker, *GTFL-1* that was linked to the gynoecious locus at a distance of 5.46 cM by using restriction-associated DNA tag sequencing (RAD-seq) analysis. Bitter gourd draft genome sequence (Urasaki et al. 2017) of a monoecious inbred line, OHB3-1, was analyzed through illumina sequencing and de novo assembly; scaffolds of 285.5 Mb in length were generated corresponding to $\sim 84\%$ of the estimated genome size of bitter gourd (339 Mb). Draft genome sequence of bitter gourd revealed that the MOMC3_649 in bitter gourd was presumed to be an ortholog of CmAcs11 (female flower determination in melon) and two proteins (MOMC46_189, MOMC518_1) were found in bitter gourd similar to CmAcs-7 (unisexual flower development in melon) grouped in the same clade in the phylogenetic tree sequence (Urasaki et al., 2017). Cui et al. (2018) developed the RAD-based genetic map for anchoring scaffold sequences and identified quantitative trait loci (QTLs) for gynoecy, first flower node, female flower number, fruit epidermal structure, and fruit color in bitter gourd.

Elshire et al. (2011) have developed simple and highly multiplexed genotyping by sequencing (GBS) approach for population studies, germplasm characterization, and mapping of desired traits in diverse organisms. The consensus of read clusters across sequence-tagged sites becomes the reference in case of crops that lack reference genome sequence. The innovative GBS approach offers an ultimate marker-assisted selection (MAS) tool to accelerate crop improvement program (He et al. 2014). A high-density and high-resolution genetic map was constructed in bitter gourd by Gangadhara Rao et al. (2018), and a total of 2013 high-quality SNP markers binned to 20 linkage groups (LG) spanning a cumulative distance of 2329.2 cM were developed. Individual LGs ranged from 185.2 cM (LG-12) to 46.2 cM (LG-17) and average LG span was 116.46 cM. The number of SNP markers mapped in each LG varied from 23 markers (in LG-20) to 146 markers (in LG-1) with an average of 100.65 SNPs per LG. The average distance between markers was 1.16 cM across 20 LGs and average distance between the markers on the LGs ranged from 0.70 (LG-4) to 2.92 (LG-20). A total of 22 QTLs for four traits (gynoecy, sex ratio, node, and days at first female flower appearance) were identified and mapped on 20 LGs. The gynoecious (gy-1) locus is flanked by markers TP_54865 and TP_54890 on LG 12 at a distance of 3.04 cM to TP_54890 and the major QTLs identified for the earliness traits will be extremely useful in marker development and MAS for rapid development of various gynoecious lines with different genetic background of best combiner for development of early and high yielding hybrids in bitter gourd.

Genomic technologies now facilitate the rapid and cost-effective assembly for several crop genomes, the analysis of large populations of crop wild relatives, and the generation of excellent functional genomics resources for some crops. Bitter gourd being the important medicinal and vegetable crop, the genomic tools could be used to identify genetic variation underlying phenotypes of relevance for improvement in this crop, as well as to access a wider range of genetic variation in wild relatives of this crop. Most of the economically important traits related to the fruit yield, flowering, and plant type are poorly understood. There is an urgent need to develop high throughput markers (SNPs) closely related to these traits. This crop is well known for its medicinal properties. The molecules responsible for different health benefits need to be characterized systematically and development of molecular markers and understanding the biochemical pathways for important phytomedicinal traits will enable to develop bitter gourd genotypes with higher concentration of these compounds. Besides, resistance to important disease pests like mildews, leaf spot, leaf curl virus, fruit fly, and nematodes need to investigated extensively. Understanding the mechanisms for tolerance to important abiotic stresses like drought, heat, salinity and mineral toxicity and QTL mapping for these traits can be undertaken to develop climate smart genotypes in bitter gourd. Studies on omics for important traits can facilitate the bitter gourd improvement program in a great way. There is a need to develop heterotic pool and genomic database with the availability of cost-effective and efficient sequencing technology. Moreover, initiating research on genome editing to improve few targeted economically important traits can facilitate the development of bitter gourd genotype for the future.

References

Basch E, Garbardi S, Ulbricht C (2003) Bitter melon (*Momordica charantia*): a review of efficacy and safety. Amer J Health-Syst Pharm 60:356–359

Behera TK, Behera S, Bharathi LK, Joseph JK (2010) Bitter gourd: botany, horticulture and breeding. In: Janick J (ed) Horticulture reviews. Wiley, Blackwell, pp 101–141

Behera TK, Dey SS, Munshi AD, Gaikwad AB, Pal A, Singh I (2009) Sex inheritance and development of gynoecious hybrids in bitter gourd (*Momordica charantia* L.). Sci Hort 120:130–133

Behera TK, Dey SS, Sirohi PS (2006) DBGy-201 and DBGy-202: two gynoecious lines in bitter gourd (*Momordica charantia* L.) isolated from indigenous source. Indian J Genet 66:61–62

Chakravarty HL (1990) Cucurbits of India and their role in the development of vegetable crops. In: Bates DM, Robinson RW, Jeffrey C (eds) Biology and utilization of cucurbitaceae. Cornell University Press, Ithaca, NY, pp 325–334

Cui J, Luo S, Niu Y, Huang R, Wen Q, Su J et al (2018) A RAD-based genetic map for anchoring scaffold sequences and identifying QTLs in bitter gourd (*Momordica charantia*). Front Plant Sci 9:477. https://doi.org/10.3389/fpls.2018.00477

Decker-Walters DS (1999) Cucurbits, sanskrit, and the Indo-Aryas. Econ Bot 53:98–112

Dey SS, Singh AK, Chandel D, Behera TK (2006) Genetic diversity of bitter gourd (*Momordica charantia* L.) genotypes revealed by RAPD markers and agronomic traits. Sci Hort 109:21–28

Elshire RJ, Glaubitz JC, Sun Q, Poland JA, Kawamoto K, Buckler ES et al (2011) A robust, simple genotyping-by-sequencing (GBS) approach for high diversity species. PLoS ONE 6:e19379. https://doi.org/10.1371/journal.pone.0019379

Gaikwad AB, Behera TK, Singh AK, Chandel D, Karihaloo JL, Staub JE (2008) AFLP analysis provides strategies for improvement of bitter gourd (*Momordica charantia* L.). HortScience 43:127–133

Gangadhara Rao P, Behera TK, Gaikwad AB, Munshi AD, Jat GS, Boopalakrishnan G (2018) Mapping and QTL analysis of gynoecy and earliness in bitter gourd (*Momordica charantia* L.) using genotyping-by-sequencing (GBS) technology. Front Plant Sci 9:1555. https://doi.org/10.3389/fpls.2018.01555

Grover JK, Yadav SP (2004) Pharmacological actions and potential uses of *Momordica charantia*: a review. J Ethnopharmacol 93:123–132

Grover JK, Rathi SS, Vats V (2002) Amelioration of experimental diabetic neuropathy and gastropathy in rats following oral administration of plant (*Eugenia jambolana*, *Mucuna pruriens* and *Tinospora cordifolia*) extracts. Indian J Exp Biol 40:273–276

He J, Zhao X, Laroche A, Lu ZX, Liu H, Li Z (2014) Geno typing by—sequencing (GBS), an ultimate marker-assisted selection (MAS) tool to accelerate plant breeding. Front Plant Sci 5:484. https://doi.org/10.3389/fpls.2014.00484

Heiser CB (1979) The gourd book. University of Oklahoma Press, Norman, OK

Iwamoto E, Ishida T (2005) Bisexual flower induction by the application of silver nitrate in gynoecious balsam pear (*Momordica charantia* L.). Hort Res (Japan) 4:391–395

Jha UC, Roy RP (1989) Hermaphrodite flowers in dioecious *Momordica dioica* Roxb. Curr Sci 58:1249–1250

Kajale MD (1991) Current status of Indian palaeoethnobotany: introduced and indigenous food plants with a discussion of the historical and evolutionary development of India agriculture and agricultural systems in general. In: Renfrew JM (ed) New light on early farming. Edinburgh University Press, Edinburgh, UK, Recent Developments in Palaeoethnobotany, pp 155–189

Kole C, Bode AO, Kole P, Rao VK, Bajpai A, Backiyarani S (2012) The first genetic map and positions of major fruit trait loci of bitter melon (*Momordica charantia*). J Plant Sci Mol Breed 1:1–6. https://doi.org/10.7243/2050-2389-1-1

Kole C, Kole P, Randunu KM, Choudhary P, Podila R, Ke PC, Rao AM, Marcus RK (2013) Nanobiotechnology can boost crop production and quality: first evidence from increased plant biomass, fruit yield and phytomedicine content in bitter melon (*Momordica charantia*). BMC Biotechnol 26(13):37. https://doi.org/10.1186/1472-6750-13-37

Marr KL, Xia YM, Bhattarai NK (2004) Allozyme, morphological and nutritional analysis bearing on the domestication of *Momordica charantia* L. (Cucurbitaceae). Econ Bot 58:435–455

Matsumura H, Miyagi N, Taniai N, Fukushima M, Tarora K, Shudo A (2014) Mapping of the gynoecy in bitter gourd (*Momordica charantia*) using RAD-Seq analysis. PLoS ONE 9:e87138. https://doi.org/10.1371/journal.pone.0087138

Miniraj N, Prasanna KP, Peter KV (1993) Bitter gourd (*Momordica* spp.). In: Kalloo G, Bergh BO (eds) Genetic improvement of vegetable crops. Pergamon Press, Oxford, pp 239–246

Ram D, Kumar S, Banerjee MK, Kalloo G (2002) Occurrence, identification and preliminary characterization of gynoecism in bitter gourd (*Momordica charantia* L.). Indian J Agri Sci 72:348–349

Raman A, Lau C (1996) Anti-diabetic properties and phytochemistry of *Momordica charantia* L. (Cucurbitaceae). Phytomedicine 2:349–362

Reyes MEC, Gildemacher BH, Jansen GJ (1994) *Momordica* L. In: Siemonsma JS, Piluek K (eds) Plant resources of South-East Asia: vegetables. Pudoc Scientific Publishers, Wageningen, the Netherlands, pp 206–210

Roy SK (1973) A simple and rapid method for estimation of total carotenoids pigments in mango. J Food Sci Technol 10:45

Sathishsekar D, Subramanian S (2005) Antioxidant properties of *Momordica Charantia* (bitter gourd) seeds on streptozotocin induced diabetic rats. Asia Pac J Clin Nutr 14(2):153–158

Semiz A, Sen A (2007) Antioxidant and chemoprotective properties of *Momordica charantia* L. (bitter melon) fruit extract. Afr J Biotechnol 6(3):273–277

Singh AK, Behera TK, Chandel D, Sharma P, Singh NK (2007) Assessing genetic relationships among bitter gourd (*Momordica charantia* L.) accessions using inter simple sequence repeat (ISSR) markers. J Hort Sci Biotechnol 82:217–222

Sirohi PS (1997) Improvement in cucurbit vegetables. Indian Hort 42:64–67

Tindall HD (1983) Vegetables in the tropics. Macmillan, London

Urasaki N, Takagi H, Natsume S, Uemura A, Taniai N, Miyagi N et al (2017) Draft genome sequence of bitter gourd (*Momordica charantia*), a vegetable and medicinal plant in tropical and subtropical regions. DNA Res 24(1):51–58. https://doi.org/10.1093/dnares/dsw047

Walters TW, Decker-Walters DS (1988) Balsam-pear (*Momordica charantia*, Cucurbita- ceae). Econ Bot 42(2):286–292

Wang QM, Zeng GW (1997) Hormonal regulation of sex differentiation on *Momordica charantia* L. J Zhejiang Agri Univ 23:551–556

Wang Z, Xiang C (2013) Genetic mapping of QTLs for horticulture traits in a F2-3 population of bitter gourd (*Momordica charantia* L.). Euphytica 193:235–250. https://doi.org/10.1007/s10681-013-0932-0

Yang SL, Walters TW (1992) Ethnobotany and the economic role of the Cucurbitaceae of China. Econ Bot 46:349–367

Yonemori S, Fujieda K (1985) Sex expression in *Momordica charantia* L. Sci Bull Coll Agri, Univ Ryukyus, Okinawa 32:183–187

Botanical Description of Bitter Gourd

2

A. C. Asna, Jiji Joseph and K. Joseph John

Abstract

Bitter gourd (*Momordica charantia*) is one of the world's major vegetable crops, which belongs to the family Cucurbitaceae. The genus *Momordica* is a native of the Paleotropics and comprises about 60 species. Bitter gourd grows in tropical and subtropical areas, including parts of East Africa, Asia, the Caribbean, and South America, where it is used not only as a food but also as a medicine. Two botanical varieties *viz.*, var. *charantia* synonymous with large-fruited cultivated Chinese bitter melon and var. *muricata* representing small-fruited, predominantly wild forms were recognized. Wide variability was noticed especially among cultivated types for fruit and seed morphology. The plant is monoecious, annual climber with long-stalked leaves and yellow, solitary male and female flowers borne on the leaf axils. The warty and oblong or elliptical-shaped fruit is botanically a 'pepo.' The plant grows well in a variety of soils and begins flowering about one month after planting. It is used as a food, bitter flavoring, and medicine. Bitter gourd has a relatively high nutritional value due to high iron and ascorbic acid content. Indians have traditionally used the leaves and fruits as a medicine to treat diabetes, colic, and to heal skin sores and wounds. Bitter gourd is reported to possess antioxidant, antimicrobial, antiviral, and antidiabetic properties.

2.1 Introduction

Bitter gourd/bitter melon/balsam pear, *Momordica charantia* L. ($2n = 2x = 22$), belonging to the family Cucurbitaceae, is an important commercial vegetable crop grown in India, Sri Lanka, Philippines, Thailand, Malaysia, China, Japan, Australia, tropical Africa, South America, and the Caribbean. The genus *Momordica* is indigenous to the Paleotropics (Robinson and Decker- Walters 1999). The term '*Momordica*' is believed to have been derived from the Latin word '*mordeo*' which means 'to bite,' referring to the bitten appearance of the grooved edges of its seeds (Durry 1864) or the jagged edges of the leaves (Krishnendu and Nandini 2016), which appear as if they have been bitten. The word '*charantia*' is from the ancient Greek for 'beautiful flower.' Rheede's (1688) *Hortus Malabaricus* is the first-ever printed record on the descriptions and illustrations of paval (*M. charantia*). He described four entities of *Momordica* (*paval, pandipaval, erumapaval,* and *bempaval*), which formed the basis for Linnaeus and subsequent botanists to

A. C. Asna (✉) · J. Joseph
Department of Plant Breeding and Genetics, Kerala Agricultural University, Thrissur, Kerala, India
e-mail: asna.ac@gmail.com

K. Joseph John
ICAR-NBPGR Regional Station, Thrissur, Kerala, India

C. Kole et al. (eds.), *The Bitter Gourd Genome*, Compendium of Plant Genomes,
https://doi.org/10.1007/978-3-030-15062-4_2

describe this genus and some of the species (Joseph 2005). Although *Momordica* is one of the largest genera in the family Cucurbitaceae, *M. charantia* is the widely cultivated species in this genus which has been extensively studied. This vegetable has been highly valued for its nutritive and medicinal properties. Bitter gourd has been in use for centuries in the traditional system of medicine in India, China, Africa, and Latin America. The fruits are known to possess antioxidant, antimicrobial, and antidiabetic properties (Raman and Lau 1996).

2.2 Origin and Distribution

Momordica is a monophyletic genus that originated in tropical Africa and the origin of Asian species was considered as the result of one long-distance dispersal event that occurred about 19 million years ago (Schaefer et al. 2009; Schaefer and Renner 2010). In the African savanna region, the monoecious species are said to have evolved from dioecious species seven times independently (Bharathi et al. 2010). The natural geographic distribution of wild edible *Momordica* was investigated and characteristics of distribution in India were analyzed based on a study of herbarium sheets in major Indian herbaria and passport data of germplasm collections (Joseph 2005).

M. charantia was believed to be of Asian origin till a recent study using plastid and mitochondrial DNA-based markers revealed that this species is most likely of African origin (Schaefer and Renner 2010). It has a long history of

cultivation as a food and medicinal plant in Africa and Asia (Morton 1967; Walters and Decker-Walters 1988). Based on Rheede's description of the taxa in his *Hortus Malabaricus* (Rheede 1688), bitter gourd was originally described by Linnaeus (1753) from Peninsular India. Even though the original place of domestication of bitter gourd is unclear, the assumed areas proposed by various workers include southern China, eastern India (Sands 1928; Degner 1947; Walters and Decker-Walters 1988; Raj et al. 1993; Robinson and Decker-Walters 1999; Marr et al. 2004), and southwestern India (Joseph 2005). From Africa, bitter gourd was believed to have been taken to Brazil via slave trade and then to 'Middle America' (Ames 1939).

Indo-Aryans of 2000–200 BC also valued this vegetable crop as food, medicine, containers, musical instruments, and literary metaphor (Decker-Walters 1999). Even Dravidians were aware of this plant (Decker-Walters 1999). It is now found naturalized in almost all tropical and subtropical regions. It is an important market vegetable in southern and eastern Asia like India, Sri Lanka, Vietnam, Thailand, Malaysia, the Philippines, and southern China (https://uses.plantnet-project.org/en/Momordica_charantia_ (PROTA)).

2.3 Taxonomy

2.3.1 Taxonomy Tree

According to Cronquist (1988), taxonomic position of bitter gourd is as follows:

Kingdom – Plantae

Phylum – Magnoliophyta

Class: Magnoliopsida

Order – Cucurbitales

Family – Cucurbitaceae

Subfamily– Cucurbitoideae

Tribe– Joliffieae

Subtribe– Thalidianthinae

Genus– Momordica

Species– charantia

Binomial name –*Momordica charantia* L

Table 2.1 List of Asiatic species of different sects

Sl. No.	Sects	Species
1.	*Cochinchinensis*	*M. cochinchinensis* *M. dioica* *M. sahyadrica* *M. denticulata* *M. denudata* *M. clarkeana* *M. subangulata*
2.	*Momordica*	*M. charantia* *M. balsamina*
3.	*Raphanocarpus*	*M. cymbalaria*

2.3.2 Classification

The genus *Momordica*, belonging to the subtribe Thalidianthinae, subfamily Cucurbitoideae of Cucurbitaceae, comprises 60 species, of which 47 are found in Africa and 13 in southeast Asia (De Wilde and Duyfies 2002; Schaefer and Renner 2011). The genus can be divided into 11 clades (Schaefer and Renner 2010) that mostly correspond to the morphological clades proposed by Jeffrey and de Wilde (2006). The Asiatic species falls under three sects (Table 2.1). Dioecious species grouped under the sect. *Cochinchinensis* and monoecious species under the sect. *Momordica* (Schaefer and Renner 2010).

Generic and species descriptions vary in various floras published in India before 1947. Chakravarthy (1982) enumerated seven species from India including *M. denudata* from Kerala and *M. macrophylla* from the Assam–Manipur belt bordering Myanmar. He also described *M. charantia* var. *muricata* based on *Hortus Malabaricus*. *M. charantia* consists of two botanical varieties *viz.*, *M. charantia* var. *muricata*, a wild variety with small and round fruits having markedly sculptured seeds and *M. charantia* var. *charantia*, which produces large fusiform fruits having feebly sculptured seeds (Chakravarty 1990). The wild variety (*M. charantia* var. *muricata*) is considered as the progenitor of the cultivated *M. charantia* var. *charantia* (Degner 1947). Jeffrey (1980) ruled out the occurrence of

M. subangulata from India for the absence of ridged or longitudinally alate fruits and treated this component under *M. dioica*. However, Soyimchiten et al. (2015) reported its occurrence in Nagaland. Kumar and Pandey (2002) also worked on the taxonomy and diversity of the genus in India.

Joseph and Antony (2010) presented a taxonomic revision of the genus in India and recognized six species *viz.*, *M. balsamina*, *M. charantia*, *M. dioica*, *M. sahyadrica*, *M. subangulata*, and *M. cochinchinensis*. They considered the occurrence/existence of *M. denudata* in India as fairly doubtful. Later, Renner and Pandey (2013) included two more species under this genus *viz. M. denudata* and *M. cymbalaria*, which awaits proper authentication. *Momordica* species of India include three monoecious taxa *viz. M. charantia*, *M. balsamina*, and *M. cymbalaria*. The dioecious taxa are *M. dioica*, *M. sahyadrica*, *M. cochinchinensis*, *M. subangulata* subsp. *renigera.*, and *M. subangulata* subsp. *subangulata*. Taxonomic key to identify these species of India is given in Table 2.2.

Yang and Walters (1992) classified bitter gourd into three horticultural groups or types:

1. A small-fruited type where fruits are 10–20 cm long, 0.1–0.3 kg in weight, usually dark green, and very bitter
2. A long-fruited type (most commonly grown commercially in China) where fruits are 30–60 cm long, 0.2–0.6 kg in weight, light

Table 2.2 Taxonomic key to identify *Momordica* species of India

Sl. No.	*Momordica* species	Taxonomic key
I.		*Germination epigeal, annual, tap root non-tuberous, plant monoecious, nectary in male flowers not closed with corolla scales, fruits muricate, or tubercled*
a.	M. charantia	Bracts of male flowers about the middle of the flower stalk; fruits small or large, softly tubercled or muricate with long green ridges; seeds thick, flat on surface, margins edged, thick on sides, broadly rectangular, no distinction between chalazal and micropylar ends, ends subtridentate, heavily or feebly sculptured
b.	M. balsamina	Bracts of male flowers at the apex of the peduncle, fruits small, distantly soft tubercled, no bumps or ridges; seeds very thin, sides not thick, margins wedged, broadly ovate round with tapering micropylar end, ends roundish, finely pitted and feebly sculptured
II.		*Germination hypogeal, perennial, tap root tuberous, plant dioecious, nectary of the male flowers closed with prominent corolla scales, fruits echinate*
a.		Petals (three inner) with black purple blotch, male calyx hypanthium saucer shaped
i.	M. subangulata subsp. *renigera*	Robust habit, leaf cordate, unlobed, margins dentate, petiole eglandular, sepal apex acuminate, flower mostly creamish-yellow colored, male calyx blackish purple, broad, tip round-oval, fruits faintly ridged, softly profusely echinate, seeds medium sized, rectangularly cog wheel shaped
ii.	M. subangulata subsp. *subangulata*	Slender habit, leaf ovate-reniform, unlobed, glands on leaf blade margins absent, petiole eglandular, sepal apex retuse, flower deep yellow colored, male calyx saucer shaped, purplish black, fruits are distinctly non-echinate and having a close resemblance to bitter gourd with irregularly crested ribs or with broken ridges primarily along five longitudinal lines, seeds medium sized, ovoid/oblong/globose shaped and slightly sculptured (Source: Soyimchiten et al. 2015)
iii.	M. cochinchinensis	Leaf unlobed or deeply lobed, margins undulate, petiole with 3–5 prominent bead-like projections, male calyx blackish purple, broad, tip triangular, fruits highly echinate, seeds large, penta-hexagonal, subtridentate on ends
b.		Petals without purple blotch, male calyx hypanthium cup shaped
i.	M. sahyadrica	Anthesis in the early morning, flowers large, showy, bright yellow, not scented, male calyx blackish purple, sepals of male flower broad, tip oval, round, or scarious
ii.	M. dioica	Anthesis in the evening, flowers small, pale yellow, intensely musky scented, male calyx whitish yellow, sepals of male flower narrow acute
III.	M. cymbalaria	Germination hypogeal, perennial, tap root tuberous, plant monoecious, male flowers borne in short raceme, anthers asymmetrical, fruits ribbed, arils white, epicarp papery and smooth and seeds shiny, round, non-bitten.

Source Bharathi and Joseph (2013)

green in color with medium-size protuberances, and only slightly bitter

3. A triangular-fruited type where cone-shaped fruits are 8–12 cm long, 0.3–0.6 kg in weight, light to dark green with prominent tubercles, and moderately to strongly bitter.

However, being cross-pollinated and interbreeding, maintenance of these distinct types needs isolation or selfing.

Reyes et al. (1994) reclassified Indian and southeast Asian *M. charantia* botanical varieties based on fruit diameter as *M. charantia* var.

Fig. 2.1 Long viny growth habit of wild bitter gourd

Fig. 2.2 Medium viny growth habit of cultivated bitter gourd

minima Williams and Ng < 5 cm and *M. chavantia* var. *maxima* Williams and Ng > 5 cm.

2.4 Morphology

2.4.1 Habit

Bitter gourd is an annual, monoecious, and slender climbing or trailing herb, 2–4 m high, scarcely to densely pubescent. Even though the plant grows as an annual crop, it can extend beyond one season, especially wild *muricata* varieties. Medium viny nature is present in most of the cultivated varieties (Fig. 2.2), whereas the wild cultivars exhibit long viny growth habit in which the vines spread like a carpet and easily cover the support (Fig. 2.1) (Asna 2018).

2.4.2 Seedlings

Bitter gourd is an annual climber with epigeal germination where the cotyledons are brought above the ground due to the elongation of hypocotyl (Fig. 2.3). Length of hypocotyl ranged from 5.48 to 13.26 cm and that of epicotyl ranged from 1.23 to 4.67 cm. Length of hypocotyl is higher than epicotyl in all the varieties of bitter gourd (Asna 2018). When compared to wild/semi-domesticated genotypes, epicotyl and hypocotyl length was higher in cultivated types. Robustness and size of the cotyledons, leaves, and stem of the seedlings were greater in *M. charantia* var. *charantia* and progressively reduced to *M. charantia* var. *muricata.*

Fig. 2.3 Germination
behavior of bitter gourd

Fig. 2.4 Leaf characters of
bitter gourd seedlings,
a *M. charantia* var. *charantia*,
b *M. charntia* var. *muricata*

The primary leaf shape and leaf margin are uniform in all the varieties of bitter gourd, reniform, and serrate, respectively. Wild cultivars of bitter gourd produce dark green-colored leaves, whereas cultivated varieties develop green to light green-colored leaves (Fig. 2.4).

2.4.3 Root

The plant has a well-developed non-tuberous tap root which becomes thick (https://portal.wiktrop. org/species/show/216). The multilateral of the main root can reach up to 30–50 cm underground. Since bitter gourd can also be used as a transplanted crop, the root system has a very great role in the establishment of the crop. Proper care should be taken while transplanting.

2.4.4 Stem

Momordica charantia is an annual monoecious climber with medium-sized vines. Angular-shaped

Fig. 2.5 Variability in twining tendency of bitter gourd

Slight

Intermediate

stem is scarce to densely pubescent, but the tender parts are wooly in nature. Twining tendency varies from slight to intermediate depending on the cultivars (Fig. 2.5). Thin stem is present in wild/semi-domesticated cultivars, whereas stem thickness is high in the cultivated types (Asna 2018). This difference is easily visible in the seedling stage itself. Simple, unbranched, puberulous, and delicate tendrils of 12–15 cm long are present on the vines. However, in some wild varieties of bitter gourd, bifid tendrils are also observed (Bharathi and John 2013).

Length of internode is measured as the distance between the fourth and fifth nodes at full foliage stage. Internodal length was found to be in the range of 1.46–4.20 cm (Asna 2018). Sidhu and Pathak (2016) recorded internodal length in the range of 1.0 to 3.5 cm in 22 genotypes of bitter gourd evaluated. This length is found to be more in the cultivated varieties than in wild/semi-domesticated genotypes of bitter gourd. Hence, the number of nodes per plant is more and close in *muricata* types, which gives it a more compact mat-like growth habit.

2.4.5 Leaves

The plant presents characteristic leaves with serrate margins. Simple and alternate leaves of this plant are reniform to orbicular or suborbicular in outline, 2.5–8 × 4–10 cm, cordate at base, acute or acuminate at apex. Leaf blades are usually deeply and palmately 5–9 lobed. Leaf petioles are slender, glabrous, and 1.5–5 cm long. Leaf margin and leaf shape used to be uniform in all the genotypes, multifid, and cordate, respectively, as reported by Sidhu and Pathak (2016). But the variability can be seen in the leaf morphology of wild/semi domesticated types (Fig. 2.6A) (Asna 2018). Leaf size and leaf pubescence varied among the cultivars

(A)

(B)

Fig. 2.6 **A** Variability in leaf morphology of wild/semi-domesticated bitter gourd genotypes; **B** Variability in leaf size of bitter gourd genotypes, **a**. *M. charantia* var. *charantia* **b**. intermediate types, **c**. *M. charantia* var. *muricata*

Fig. 2.7 **a** Staminate flower of bitter gourd **b** Pistillate flowers of bitter gourd

(Fig. 2.6B). Wild/semi-domesticated bitter gourd accessions usually exhibit smaller leaves with intermediate pubescence compared to the larger leaves of the cultivated varieties (Fig. 2.6).

2.4.6 Inflorescence/Flowers

Bitter gourd is monoecious, where staminate (Fig. 2.7a) and pistillate flowers (Fig. 2.7b) are borne on separate nodes. The yellow color and scent of the flower attract pollinators. The male flowers are solitary, pale yellow, and sweet-scented. They are incomplete, imperfect, regular, actinomorphic, apopetalous, and whorled. Stalk is slender with a green-colored reniform-shaped bract of about 5–11 mm diameter at midway or toward base. The portion above bract is pedicel and below bract is peduncle. Peduncle and pedicel are approximately 2–5 and 2–6 cm long, respectively. The petals ($n = 5$) and the sepals ($n = 5$) are free. Sepals are ovate-elliptic, $4–6 \times 2–3$ mm, and pale green in color touching each other and protecting the corolla tube. Petals are obovate, $10–20 \times 7–15$ mm, mucronate at apex, with two scales. There are five stamens but there appeared to be three; two sets have two fused anthers, producing two compound stamens and one simple stamen. They are bell-shaped and 4 mm long. Filaments are three, two are bilocular and one is unilocular; 1.5–2 mm long, inserted at nectary (Behera et al. 2010). The anthers are yellow, coherent, and dorsifixed. Disk is shortly cup shaped.

Female flowers are easily recognizable by the presence of their ovaries. Perianth parts are the same as the male flowers but have a stronger scent than the male flowers. Female flower peduncle is 1–6 cm long; bract 1–9 mm diameter; pedicel 1–8 cm long: sepals are narrow, oblong lanceolate, and 2–5 mm long; petals are smaller than or equal to that in male flowers, 7–10 mm long. The pistillate flower of bitter gourd consists of an inferior (epigynous), fusiform, narrowly rostrate, $5–11 \times 2–3$ mm, muricate, tuberculate or longitudinally ridged ovary and a three-lobed, wet stigma that is attached to a columnar, hollow style (Pillai et al. 1978). There are three short styles of 2 mm length terminated by three bilobed or divided stigmas. The receptive stigma is moist and bright green. The ovary contains three fused carpels (syncarpous), each with 14–18 ovules, surrounded by an ovary wall. Although the number of ovules in an ovary can be up to 60, the average is 40. Anatropous ovules are attached to parietal placenta in two irregularly aligned rows in each carpel. Not more than four ovules can be seen in ovary cross section (Behera et al. 2010).

Even though the most prominent sex form in bitter gourd is monoecious, gynoecious sex form has been reported from India, Japan, and China (Zhou et al. 1998; Ram et al. 2002; Iwamoto and Ishida 2006). The generations advanced using these gynoecious lines as one of the parents showed very high percent of pistillate flowers and thereby higher yield (Ram et al. 2002). Dey et al. (2010) identified a gynoecious line, DBGy-201, and proved it as a good combiner with high general combining ability (GCA) and specific combining ability (SCA) effects. Similarly, the gynoecious lines can also be advanced and can be used more effectively in bitter gourd improvement programs especially for yield and earliness.

2.4.7 Fruit

The bitter gourd varieties differ substantially in the shapes and bitterness of the fruit. Fruits are pendulous with 2–8 cm long stalks. They are discoid, ovoid, ellipsoid to oblong or blocky in shape, often narrowed at ends, sometimes finely rostrate, $3–20 \times 2–5$ cm. The fruit has a tuberculate surface with 8–10 broken or continuous ridges. It is emerald-green, whitish green, or rarely white in color when immature, turning orange yellow during ripening. It is hollow in cross section, with a relatively thin layer of flesh surrounding a central seed cavity filled with seeds and pith. At the mature stage, it exposes red seeds, surrounded by a slimy aril, splitting

Fig. 2.8 Variability in fruit morphology of bitter gourd

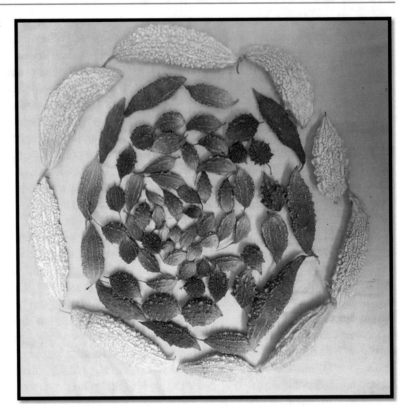

from base into three irregular valves. The qualitative characters of fruits like fruit shape, fruit ends, fruit color, fruit luster, fruit surface ornamentation, nature and density of tubercles, etc. used to show wide variability among the cultivars (Fig. 2.8). The fruit morphology varied greatly in color, size, and exocarp characteristics (Behera et al. 2008). For instance, Indian cultivars belonging to var. *charantia* bear large fusiform fruits while wild *muricata* ecotypes develop small and round fruits (Chakravarty 1990). Even though *M. charantia* var. *muricata* has close resemblance to *M. charantia*var. *charantia*, they can be easily identified by the smaller size of fruits and seeds (Bharathi and John 2013).

The color of mature fruits can be light green, green, dark green, or a mixed color combination with green on one side and light green color on the other side. The cultivated varieties exhibit the typical whitish green color, which is the characteristic feature of var. *charantia*. The mature fruits belonging to different cultivars can be matt or intermediate or glossy with respect to their fruit luster.

A wide range of genetic diversity existed in bitter gourd with respect to fruit morphology and surface texture (Robinson and Decker-Walters 1999). Since morphological characters are the primary tool for the identification of a genotype and for establishing phylogenetic relationships, fruit morphology has a great role in the identification of various *Momordica* spp. (Sidhu and Pathak 2016). The fruit shape of bitter gourd can be disk, rhomboid, cylindrical, spindle shaped, elliptical, oblong, or globular (NBPGR 2001; Fig. 2.9). The fruit size of *M. charantia* var. *muricata* is very small when compared to *M. charantia* var. *charantia* and many natural intermediate types were also known (Njoroge and Luijk 2004).

Fruit ends (Fig. 2.10) are normally pointed either at both ends or only at blossom end. In certain cultivars, both ends can be round (blossom end shape is blunt). Generally, the fruit ribs are

Fig. 2.9 Variability in shape of bitter gourd fruits

broken in South Indian bitter gourd genotypes. But, Pusa Do Mausmi (*M. charantia* var. *charantia*) from IARI is a variety with continuous ribs (Bharathi et al. 2012) (Fig. 2.11). Chance mating between this variety and the wild types might result in the appearance of continuous ribs in the wild/semi-domesticated genotypes. Fruit surface also shows wide variability among the

Both ends pointed   **Only blossom end pointed**   **Both ends round**

Fig. 2.10 Variability in ends of bitter gourd fruits

Fig. 2.11 Variability in ribs
of bitter gourd fruits

Broken ribs   **Continuous ribs**

genotypes. The surface of bitter gourd fruits is highly tubercled with long green ridges. Surface ornamentation can be recorded in cross section at marketable stage as light or deep tubercle. Usually, sharp-pointed tubercles and soft raised tubercles are found in wild/semi-domesticated and cultivated varieties, respectively.

The number of fruits per plant and fruit yield per plant ranges from 14.32 to 168.39 and 124.05 g to 8568.61 g, respectively (Asna 2018). Fruit yield is the direct contribution of individual fruit weight and number of branches per plant (Islam et al. 2014). It is observed that the small fruit bearing wild/semi-domesticated genotypes produced larger number of fruits per plant. It might be due to the presence of larger number of primary and secondary branches in these wild genotypes (Islam et al. 2014). A positive linear relationship is present between the number of pistillate flowers and fruits per plant. As the number of female flowers increases, the number of fruits produced also increases. Both these yield contributing characters are the highest in the cultivated types of bitter gourd. Majority of the wild/semi-domesticated genotypes of bitter gourd produce a greater number of pistillate flowers and fruits than the cultivated *charantia* varieties. Even though more number of fruits is produced by the *muricata* accessions, other yield contributing characters like fruit weight, length, breadth, flesh thickness, and cavity size are found to be higher in the commercial *charantia* varieties. Cultivated types have thick fruit wall (3–4 mm thick), whereas wild *muricata* forms have very thin (0.1– 1.5 mm) fruit wall (Marr et al. 2004).

The average weight of individual fruit varies among cultivars due to genotypic variation and ranges from 3.53 g to 420.61 g (Asna 2018). The results show that the average fruit weight, fruit length, and fruit width directly contribute to fruit yield. Cavity size varies from 0.88 to 3.12 cm with maximum value in cultivated varieties (Asna 2018). The highest flesh thickness was also associated with fruit length and fruit size to accommodate the fruit biomass (Dey et al. 2006). Length and width of fruit are lowest in wild bitter gourd (2.13 and 1.05 cm) and the highest in

cultivated types (Asna 2018). As fruit length and width increase, fruit weight also increases.

The yield is mainly contributed by fruit weight, fruit length, fruit width, flesh thickness, and cavity size. The characters like number of pistillate flowers and number of fruits per plant also contributed toward the yield but the role played by these characters is very low when compared to other fruit characters. These yield attributes are influenced by morpho-physiological characters like vine length, primary branches, leaf area, and chlorophyll content (Mia et al. 2012; Mia and Shamsuddin 2011; Ram et al. 2002). The fruit size is the direct indicator of yield increment. Therefore, major emphasis should be given on selection of genotypes having a greater number of fruits per plant, high average fruit weight, fruit length, and fruit diameter which would lead to the development of high yielding cultivars of bitter gourd.

2.4.8 Seeds

Seeds are numerous (5–30), squarish rectangular in shape with subtridentate ends, compressed and sculptured faces. Margins are grooved. Testa can be brown, black, or straw colored. Seed characters also have an inseparable role in distinguishing wild *muricata* accessions from the commercially cultivated varieties. Wild bitter gourd produces small seeds resembling cockroach eggs, whereas genotypes belonging to *charantia* develop comparatively large, broadly rectangular, heavily or feebly sculptured seeds with subtridentate ends (Bharathi and John 2013). Majority of bitter gourd genotypes have straw-colored seeds. In addition to this, black, black, and brown patched, brownish tan, and whitish brown-colored seeds are also noted especially in wild and semi-domesticated forms. Seeds may be glossy or matt. Seeds with intermediate luster are also noted in certain genotypes. The seed characters like surface evenness, ends and sides of seeds and surface sculpturing are unique for each genotype. The seed surface is either flat and creaked or pitted or invaginated with dented or smooth sides and feebly or clearly

Fig. 2.12 Variability in seeds of bitter gourd accessions

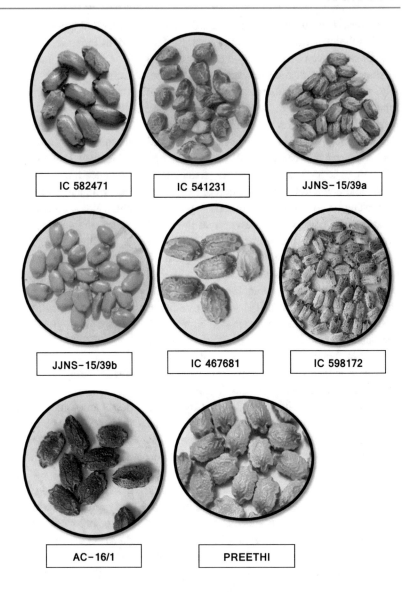

subtridendate ends. Variability in seed characters of bitter gourd is depicted in Fig. 2.12.

The quantitative characters of seeds like number of seeds per fruit, seed length, seed width, and 100 seed weight vary greatly with cultivars. The wild and cultivated genotypes can be easily identified by the seed size (Bharathi and John 2013). The number of seeds per fruit ranges from 4.69 to 24.79 (Asna 2018). It was noticed that the wild/semi-domesticated genotypes are highly seeded than the *charantia* types. Seediness is a wild character which helps in survival through the production of more offsprings.

The cultivated varieties, with long fusiform fruits and high fruit weight, produce only 19.81–21.58 seeds per fruit (Asna 2018). While the wild *muricate* types produce 2.34–9.57 seeds per fruit. The smaller size and highly seeded nature of *muricata* types may be the reason for the low acceptability of this highly medicinal variety of bitter gourd among the consumers. Reduced seed content of fruits is a desirable character in bitter gourd (Islam et al. 2009).

Weight of 100 dry seeds was also low in small-fruited genotypes ranging from 4.98 to 19.16 g due to the smaller size of the seeds

(Asna 2018). Even though the number and size of seeds of the wild/semi-domesticated genotypes were lower than that of the var. *charantia*, the proportion of the seeds to the total fruit biomass was higher. So, when the seeds are removed, the consumable portion of the fruit becomes very small in *muricata* types. Hence, these fruits are preferred for cooking at immature stage along with the seeds.

2.5 Floral Biology

In bitter gourd, the first flower appears 35–55 days after sowing (Rasco and Castillo 1990; Reyes et al. 1994). Usually, the male (staminate) flowers emerge first and the female (pistillate) flowers appear after 9–42 days but under long-day conditions, male flowers even bloom two weeks before female flowers (Palada and Chang 2003). The flowering continues for about six months in cultivated types and 8–12 months in wild genotypes (Asna 2018). Genotypes with first female flower emergence followed by male flowers after a few days were also observed (Asna 2018). The male and female flower buds used to take on an average 17–19 and 21–22 days, respectively, for their complete development (Deyto and Cervancia 2009). The anthesis occurs between 3.30 and 7.30 am (Miniraj et al. 1993) and stigma remains receptive from 24 h before to 24 h after anthesis and most receptive during early hours of the day (Rasco and Castillo 1990). The anthesis starts, before the day of opening, by the elongation of the corolla above the calyx. On the day of opening, corolla loosens up, signifying opening. Then the apexes of the petals separate, forming a bell-shaped corolla and the stamens become visible. The flowers become fully open from 6.00 a.m. to 9.55 a.m. Anther dehiscence occurs about one hour before flower opening (5.00 a. m.–7.30 a.m.). Pollen grains of bitter gourd are round shaped and shed by longitudinal splitting of anther. Female flowers open synchronously and in the same manner as that of male flowers. The male flowers drop off on the same day, while female flowers wither away next morning.

A successfully pollinated flower starts to set fruits on the second to fifth day after it opens. Unpollinated flowers dry up and the ovaries become yellow colored (Deyto and Cervancia 2009).

2.6 Sex Phenology

Flowering behavior of bitter gourd varies with cultivar, climatic conditions, and cultural practices (Deshpande et al. 1979; Rajput et al. 1996; Rasul et al. 2004; Asna 2018). In general, the staminate flowers appear at 7–17 nodes and pistillate flowers at 13–21 nodes in bitter gourd genotypes (Islam et al. 2009). In cucurbits, where inflorescence is solitary and racemose type, the staminate flowers are induced earlier in basal nodes and pistillate flowers later in proximal node (Ram et al. 2002). The flowers initiate at lower nodes in wild/semi-domesticated genotypes when compared to cultivated types (Asna 2018). This difference in flower initiation is due to genetic characters of the genotypes. The number of nodes to first female flowering is a fair measure of both sex tendency and maturity. The lower the node number, the higher the female tendency and earlier the variety (Shiefriss and Galun 1956). Genotypic and phenotypic correlations of nodal position of the first pistillate flower with lowering time and parthenocarpic yield are high (Shawaf and Baker 1981). This would help in selecting early and high yielding varieties.

The number of days required for the first staminate flower opening ranged from 39.03 to 60.16 days and pistillate flower from 40.00 to 67.06 days (Asna 2018). Days to first female flower had high negative correlation with number of fruits per plant but positive correlation with weight of fruits both at genotypic and phenotypic levels (Srivatsava and Srivatsava 1976).

Bitter gourd is a monoecious plant, naturally inducing more number of staminate flowers than the pistillate flowers, with a mean male to female ratio of 25:1 (Palada and Chang 2003). This is a common problem in bitter gourd cultivation, since this flowering behavior results in lower fruit set and yield. In order to have higher yield, the staminate and pistillate flower ratio needed to be synchronized (Ram et al. 2000). The number of flowers per plant is more in *muricata* genotypes than in cultivated varieties (Asna 2018). The flower initiation at lower nodes in wild/semi-domesticated genotypes might have played a significant role in producing more number of flowers. So, the germplasm lines having higher sex ratio than cultivated varieties have an important role in breeding programs for developing high yielding crop varieties (Dey et al. 2007).

2.7 Pollination

Bitter gourd is a cross-pollinated crop. Pollen tubes penetrate papillae tissue within 1 h of pollination arriving at the ovary cavities about 6 h after pollination, and thus, fertilization is accomplished within 18–24 h postpollination (Chang et al. 1999). In a monecious crop like bitter gourd, insect pollinators play a crucial role in pollination. The flowers of bitter gourd are borne on the leaf axils of the plant. The open position of the flowers makes it easy for the pollinators to access and exploit floral resources. The flat structure of the flower of bitter gourd is ideal for easy landing of pollinators and enhances visit by varied floral visitors. Its yellow color and scent also attract pollinators. Since the flowers are short lived, period of flower opening is very important for effective pollination. The pollinators should visit the flower during this period to affect fruit set. The size and complexity of floral parts influence the behavior of pollinators (Goulson 1999). Production of numerous male flowers relative to the female flowers enhances effective pollination through abundance of pollen grains.

Deyto and Cervancia (2009) and Subhakar et al. (2011) recorded a variety of insect pollinators on bitter gourd belonging to four orders: Hymenoptera, Lepidoptera, Coleoptera, and Diptera. The flight activity and visitation rates of pollinators are influenced by anthesis and environmental factors (Faegri and van der Pijl 1979; Forbes and Cervancia 1994; Kearns and Inouye

1993). They are inactive during high wind or heavy rain and at low temperature. The presence of pollinators is also related to nectar secretion (Cervancia and Barile 1993). Even though the visit of various insects is observed to hover around the plant, the insects of family Apidae are the most reliable agents for pollination. Within this family, honey bees are highly important as they act as active pollen carriers (Tewari and Singh 1983). The other insects are just chance pollinators.

2.8 Germination

Bitter gourd is usually grown as an annual crop, but it can also perform as a perennial in mild areas and frost-free winters (Singh and Ram 2005). Seeds are orthodox with epigeal germination. It can be either sown directly in the field or allow them to grow in a germination tray first and then transplanted at two-leaf stage. Germination of bitter gourd seed is adversely affected when temperature goes below 18 °C (Fonseka and Fonseka 2011). Seeds show thermo-dormancy on cold storage, otherwise bitter gourd does not show any dormancy. The optimum temperature for seed germination is 25 −28 °C (Peter et al. 1998). But field emergence is always a problem in bitter gourd even with high seed germinability, due to the thick seed coat. Hard seed coat slowly absorbs the water, hence the slow germination. Normally, 7–11 days are taken for seed germination. Seed soaking is a useful measure for bitter gourd growers to assure a successful seedling establishment (Lin and Sung 2001). Soaking of seeds in slightly warm water for 30 min followed by retention of seeds in a wet gunny bag or cloth bag in a warm place for 3–4 days increases the speed of germination. Poor germination percentage is common at suboptimal temperatures (Peter et al. 1998). Pre-sowing treatments such as priming (mixing seeds with moist vermiculite and keeping it for 36 h at 20 °C) and hot water (soaking seeds for 4 h in hot water at 40 °C) are therefore recommended for successful seedling establishment under suboptimal temperature (Lin and Sung 2001; Hsu et al. 2003). High seed germination was also noted by Nath et al. (1972), by soaking the seeds in 50 ppm GA$_3$ for 12 h.

2.9 Anatomy

Even though anatomical studies of Cucurbitaceae were conducted by many scientists, *viz.,* Yasuda (1999), Metcalfe and Chalk (1950), Gill and Karatela (1982), detailed study of bitter gourd including cross sections of stem, leaf, and upper and lower sections of leaves was conducted by Poyrozki and Derdovski (2017). The stems owned typical anatomical properties of a climbing dicotyledon plant. The leaves are amphistomatic and a lot of cystoliths were present on the lower surface of leaves. Stomata are anomocytic and distributed more at the lower surface of leaves. A detailed anatomical description of cross sections of immature fruits and seeds is reported for the first time by Giuliani et al. (2016).

Stem: The outermost anatomical tissue of five angled stem is epidermis, with thin and smooth cuticle. Below the cuticle, the epidermis lies with one cell line. Collenchyma cells are located only at the ridges of the stem. After 1–8 layers of sclerenchyma cells, ten bicollateral primary vascular bundles are situated in two rings which are separated by parenchymatic rays bearing prismatic crystals. Five fibro-vascular bundles, comprised of outer phloem, xylem, and inner phloem with a sclerenchyma line before the outer phloem, are present in each ring. Pith is parenchymatic (Poyrozki and Derdovski 2017).

Leaf: Leaves are dorsiventral, amphistomatic with anomocytic (or ranunculaceous type) stomata and cystoliths on the lower surface (Poyrozki and Derdovski 2017). The midrib is distinct. Upper and lower epidermises are in a line with rectangle, ovoid cells. Some collenchyma cells are present under the xylem. Leaves are dorsiventral. Palisade parenchyma cells are 1–2 layered and spongy parenchyma cells are 2–3 layered. Gill and Karatela (1982) and Metcalfe and Chalk (1950) reported the presence of glandular hairs with uniseriate stalks and spherical or disk-shaped heads on leaves.

Fruit: The fruit has a tuberculate surface with numerous swellings. It is characterized by a thin epicarp, multilayered mesocarp, and by an inconspicuous endocarp. The pericarp is characterized by five distinct tissues as follows (Giuliani et al. 2016):

1. Epicarp—the cells are generally isodiametric with those close to the stomata being tangentially elongate. The outer periclinal walls are covered by a thin cuticle and the underlying two–three layers have occasionally thickened walls.
2. Outer mesocarp—this tissue consists of 5–7 layers of large isodiametric or radially elongated cells, apparently devoid of cell sap, and a network of tiny intercellular space.
3. Middle mesocarp—the cells are tangentially elongated and thick-walled, often turgid with a watery cell sap and contain a small amount of starch. Numerous bicollateral vascular bundles occur throughout the middle mesocarp: they are either small and soft or large and stiff, forming an anastomosing vein-like system arranged in a ring.
4. Inner mesocarp—several layers of thin-walled cells, forming this tissue, present a huge amount of starch in comparison with the preceding layer. The cells are large with wide intercellular space. The inner layer bears stomata and delimits a cavity containing the seeds.
5. Endocarp—this tissue, formed by very small, thin-walled, and tangentially elongated cells, is thin and translucent.

Seed: The seeds are oblong with grooved margins and a sculptured surface. The seed coat may be differentiated into five distinct layers enclosed by the aril as follows (Giuliani et al. 2016):

1. Epidermis—this tissue consists of a single layer of prismatic palisade cells. They are generally of equal height over the flat surface of the seed, increasing in height at the sculpturations. The radial walls are often uniformly thickened; nevertheless, in some

cases, they have either straight or branched thickenings running from the inner to the outer tangential walls. The outer walls are thickened. A noteworthy feature is the presence of large starch grains in this layer.

2. Hypodermis—this tissue is made up of four layers of small, isodiametric, tightly closed sclerenchymatized cells, without intercellular space.

3. Sclerenchyma—this layer consists of exceedingly thick-walled cells. The walls are sinuous and the starch content is massive.

4. Aerenchyma—many cell layers of spongy parenchyma, differing greatly in size and shape, make up this layer. The cells of the outer layers are usually small and frequently sclerenchymatized. Underneath these small cells are one or more layers of either large or small thin-walled cells having very large intercellular spaces. The vascular bundles are embedded in this layer at the flattened surface of the seed.

5. Chlorenchyma—a single layer of small, polygonal, and inconspicuous thin-walled parenchyma cells, containing chlorophyll, forms the inner tissue.

The leaf-like cotyledons have an epidermis of small cells below which, on the inner side, are two sharply defined palisade layers. All the cells are filled with oil and protein granules. The oil in bitter gourd seeds of Indian species has a unique odor.

2.10 Ecology

2.10.1 Climate and Soil

Based on the type and extent of vine growth, the space requirement per plant varies. Other cultural needs are similar for all the *Momordica* species. Bitter gourd grows well in hot, humid areas. Bitter gourd is mainly cultivated during the spring, summer, and rainy seasons, with some winter production in subtropical climates. In contrast, it is cultivated throughout the year in tropical climates. The optimum temperature for good plant growth is 25–30 °C. Frost can kill the plants, and cool temperatures will retard development. The bitter gourd crop can grow above 18 °C (Larkcom 1991), with 24–27 °C being optimum (Desai and Musmade 1998). Bitter gourd performs well in full sun and is adaptable to a wide range of soil types but grows best in a well-drained sandy loam soil that is rich in organic matter. It grows well in soils of shallow to medium depth (50–150 cm), and like most cucurbits, bitter gourd prefers well-drained soils. For bitter gourd, the optimum soil pH is 6.0–6.7, and it can also tolerate alkaline soils up to pH 8.0, whereas spine gourd prefers a PH of 6.0–7.0.

2.11 Chemical Composition

Compared to other members of Cucurbitaceae family, bitter gourd has relatively higher nutritional value, mainly due to the iron, phosphorous, and ascorbic acid content (Oliver 1960). The nutritional value of wild bitter gourd (*M. charantia* var. *muricata*) is reported to be higher or at par than that of the cultivated ones. In addition to the medicinal properties, bitter gourd can also serve as a good source of nutrients for human health (Bakare et al. 2010). The proximate principles and nutrient compositions of bitter gourd fruits are presented in Table 2.3.

It is an excellent source of vitamins B1, B2, B3, Vitamin C, magnesium, folic acid, zinc, phosphorus, and manganese and has high dietary fiber (Keeding and Krawinkel 2006). But Vitamin A and calcium content are very low (Morton 1967). It is rich in iron and contains twice the beta-carotene of broccoli, twice the calcium of spinach, and twice the potassium of a banana (Aboa et al. 2008; Wu and Ng 2008). Nevertheless, its pods are rich in phytonutrients like dietary fiber, minerals, vitamins, and antioxidants (Klomann et al. 2010).

2.12 Economic Botany

Bitter gourd is mainly grown as a vegetable crop for its bitter tender fruits. The bitterness of bitter gourd is due to the cucurbitacin-like alkaloid

Table 2.3 Proximate principles and nutrient composition of bitter gourd (Momordico *charantia* L.) fruit

Proximate principles	Quantity
Moisture (g/100 g)	83.20
Carbohydrates (g/100 g)	10.60
Proteins (g/100 g)	2.10
Fiber (g/100 g)	1.70
Calcium (mg/100 g)	23.00
Phosphorous (mg/100 g)	38.00
Potassium (mg/100 g)	171.00
Sodium (mg/100 g)	2.40
Iron (mg/100 g)	2.00
Copper (mg/100 g)	0.19
Manganese (mg/100 g)	0.08
Zinc (mg/100 g)	0.46
Beta-carotene	126.00
Vitamin C	96.00

Source Gopalan et al. (1982). Nutritive value of Indian foods. National Institute of Nutrition, ICMR, Hyderabad

momordicine and triterpene glycosides (momordicosine K and L) (Jeffrey 1980). The fruits are cooked in many ways but most commonly used after fried, boiled, stuffed, and cooked. It can be canned, pickled, or stored as dry vegetable. The fruits are an inexpensive source of proteins and minerals and rank first among cucurbits for its nutritive value. Natural antioxidants in bitter gourd enhance food quality. Bitter gourd can also be considered as a herbal drug which is a part of various traditional systems of medicine. Traditionally, plant extracts of bitter gourd have been used in the treatment of diabetics, blood diseases, rheumatism, and asthma (Behera et al. 2009).

Chakravarty (1959) documented the ethnobotanical uses of the plant in India. Even though both varieties of bitter gourd grow abundantly throughout India, *M. charantia* var. *muricata* is very important as far as ethnomedical practices are concerned. Both varieties are used in traditional medicine for the treatment of diabetes and other stomach complaints in many countries. The immature and ripe fruits are cooked and eaten as vegetable in Asia and Africa (Bharathi and John

2013). Young shoots and leaves are also cooked and eaten as vegetable in India and used as flavoring agent in Java and Philippines (Anonymous 1952). Leaf and fruit extracts are used in the preparation of tea and are a popular health drink in Japan (Reyes et al. 1994). It has been used for centuries in ancient traditional Indian, Chinese, and African pharmacopoeia as anthelmintic, laxative, digestive stimulant, and to enhance appetite (Bharathi and John 2013). In Ayurveda, *M. charantia* is grouped under vegetable class of medicine and claimed to possess several therapeutic properties like regulation of digestion and metabolism, softening and clearing the motion, and improving digestion of sweet substances (Bharathi and John 2013). The plant is generally used as a hypoglycemic and antidiabetic agent. Traditionally, wild bitter gourd leaves are crushed to obtain juice for applying on the skin for treating insect bites, bee stings, burns, contact rashes, and wounds (Joseph and Antony 2008). The pharmacological profile of bitter gourd exhibits its potential as antidiabetic, antibacterial, antiviral, anticancer, antifertility, anti-ulcer, immunomodulator, antipsoriasis, analgesic, anti-inflammatory, hypotensive, hypocholesterolemic, antioxidant, cardioprotective, anthelmintic, and antimalarial agent.

2.12.1 Phytochemical Properties

More than 200 medicinal compounds have been isolated from the leaves, stems, pericarp, entire plants, callus tissues, and seeds of bitter gourd. These biologically dynamic chemicals include various glycosides, saponins, alkaloids, fixed oils, triterpenes, proteins, steroids, inorganic compounds, carotenoids, carbohydrates, *etc.* Various phytoconstituents, such as momorcharins, momordenol, momordicilin, momordicins, momordicinin, momordin, momordolol, charantin, charine, cryptoxanthin, cucurbitins, curbitacins, cucurbitanes, cycloartenols, diosgenin, elaeostearic acids, erythrodiol, galacturonic acids, gentisic acid, goyaglycosides, goyasaponins, and multiflorenol, have been isolated from MC (Table 2.4) (Husain et al. 1994;

Xie et al. 1998; Yuan et al. 1999; Parkash et al. 2002).

The medicinal value of bitter melon lies in the bioactive phytochemical constituents. These are the non-nutritive chemicals that produce definite physiological effects on human body and protect them from various diseases. In *M. charantia*, primary metabolites are common sugars, proteins, and chlorophyll while secondary metabolites are alkaloids, flavonoids, tannins, and so on. Phytosteroids are pharmologically important for human life. Diosgenin is synthesized in all plant parts with higher concentrations in fruit of *M. charantia*. All the phytochemicals possess hypoglycemic action. Alkaloids and saponins are present in *Momordica* and volatile components are released during cooking which enhances the flavor.

Photochemical studies revealed that this plant contains lutein and lycopene which are responsible for its antibiotic and antitumor activities, charatin, momordicine, and other alkaloids, saponins, phenolic constituents, glycosides, and 5-hydroxyl tryptamine. Antibacterial, antineoplastic, antiviral, and antimutagenic activities of the plant have also been reported. They also contain an array of biologically active proteins, namely momordin, α- and β-momorcharin, cucurbitacin, and MAP30, that have shown to have highly effective antihuman immune deficiency (HIV), antitumor, antidiabetic, and antirheumatic properties and to function as febrifuge medicine for jaundice, hepatitis, leprosy, hemorrhoids, psoriasis, snakebite, and vaginal discharge.

2.12.2 Medicinal Uses

Bitter melon is a popular medicinal fruit particularly in Asia and Africa, where many varieties are grown. Bitter gourd exhibits a very rich pharmacological profile (Table 2.5). It has shown antidiabetic, antibacterial, antiviral, anti-HIV, anti-herpes, anti-polio virus, anticancer, antifertility, anti-ulcer, immunomodulatory, antipsoriasis, analgesic and anti-inflammatory, hypotensive, antiprothrombin, hypocholesterolemic, antioxidant (Grover and Yadav 2004), anti-obesity (Sahib et al. 2012), and cardioprotective (Temitope et al. 2013) activities. It is also used for the treatment of various other pathological conditions such as dysmenorrhea, eczema, emmenagogue, galactagogue, gout, jaundice, kidney (stone), leprosy, leukorrhoea, piles, pneumonia, psoriasis, rheumatism, and scabies. It has also been documented to possess abortifacient, anthelmintic, contraceptive, antimalarial, and laxative properties (Khan and Omoloso 1998).

Table 2.4 Phytochemicals and constituents of *Momordica charantia*

Source	Phytochemicals
Plant body	Momorcharins, momordenol, momordicilin, momordicins, momordicinin, momordin, Momordolol, charantin, charine, cryptoxanthin, cucurbitins, cucurbitacins, cucurbitanes, cycloartenols, diosgenin, elaeostearic acids, erythrodiol, galacturonic acids, gentisic acid, goyaglycosides, goyasaponins, multiflorenol
Plant body	Glycosides, saponins, alkaloids, fixed oils, triterpenes, proteins, and steroids
Fruit	Momordicine, charantin, polypeptide p-insulin, ascorbigen, Amino acids—aspartic acid, serine, glutamic acid, threonine, glutamic acid, threonine, alanine, g-amino butyric acid and pipecolic acid, luteolin Fatty acids—lauric, myristic, palmitic, palmitoleic, stearic, oleic, linoleic, linolenic acid
Seed	Urease Amino acids—valine, threonine, methionine, isoleucine, leucine, phenylalanin, glutamic acid

Table 2.5 Pharmacological and phytochemical profile of bitter gourd

Sl. No.	Medicinal use	Responsible phytoconstituent	Effect
1.	Abortifacient	β-momorcharin	Inhibited the biosynthesis of cultured endometrial cells
2.	Anti-inflammatory	Momordicin Ic and its aglycone, oleanolic acid	Reduced the inflammation, biochemical markers
3.	Anti-allergic	Momorcharin	Depressed the delayed-type hypersensitivity response and the tumoral antibody formation
4.	Antimicrobial	β-sitosterol	Effective against *E. coli, Pseudomonas aeruginosa, Staphylococcus aureus, Klebsiella pneumonia*
5.	Anticancer	MAP 30	Activity against lymphoma, prostatic cancer, lymphoid leukemia, breast cancer, choriocarcinoma, Hodgkin's disease, etc.
		Charantagenins D & E	Active against hepatoma carcinoma, cell line Hep 3B, lung cancer cell line A549, and glioblastoma cell line U87
		7-oxo-stigmasta-5,25-diene-3-O-β-d-glucopyranoside	Active against hepatoma carcinoma, cell line Hep 3B, lung cancer cell line A549 and glioblastoma cell line U87
		Kuguacin J	Inhibit cell growth and proliferation and induce apoptosis
		Momordin I, Id and Ie	To inhibit the protein synthesis of human choriocarcinoma and trophoblasts
		α-Momorcharin	Anticancer potential gainst melanoma cells Choriocarcinoma
6.	Antioxidant	Phenolic content	Free radical scavenging activity
7.	Anti-ulcer	Momordin Ic	Inhibit ethanol induced gastric mucosal lesions

(continued)

Table 2.5 (continued)

Sl. No.	Medicinal use	Responsible phytoconstituent	Effect
8.	Antiviral		
	(i) Anti-HIV	MAP 30	
	(ii) Anti-herpes	MAP 30 and GAP 31	Inhibition of protein synthesis
9.	Anti-polio virus	Ribosome-inactivating protein	Inhibited poliovirus replication by inhibiting protein synthesis
10.	Immunomodulatory	Alpha- and beta-momorcharin	Shift in the kinetic parameters of the immune response
11.	Hypocho lesterolemic	Octadecatrienoic fatty acid	Plasma lipid peroxidation and erythrocyte membrane lipid
12.	Antidiabetic	p-insulin, v-insulin, polypeptide-p	Insulin like effect
		Momordicoside S *Momordicoside T*	Increased glucose clearance during intraperitoneal glucose tolerance test; increased basal metabolic rate and ß-oxidation
		Charatin	Hypoglycemic effect
		Conjugated linolenic acid, linoleic acid, conjugated linoleic acid	Intestinal GLP1 release
		Karavilagenine E and oleanolic acid	Intestinal GLP1 release
		Trehalose	Lowered postprandial blood glucose levels
		Momordin	PPAR β/activation
		9c, 11t, 13t conjugated linolenic acid	PPAR a and g activation
		Vicine	Hypoglycemic effect
		3b,7b,25-trihydroxycurcubita-5,23(E)-dien-19-al	Lowered blood glucose levels

Source Katiyar et al. (2017)

References

Aboa KA, Fred-Jaiyesimi AA, Jaiyesimi A (2008) Ethnobotanical studies of medicinal plants used in the management of diabetes mellitus in South Western Nigeria. J Ethnopharmacol 115:67–71

Ames O (1939) Economic annuals and human cultures. Botanical Museum of Harvard University, Cambridge

Anonymous (1952) The wealth of India: a dictionary of Indian raw materials and industrial products. Raw Mater 6:409–413

Asna AC (2018) Characterization and distant hybridization for biotic stress tolerance in bitter gourd (*Momordica charantia* L.). Ph.D. Thesis, Kerala Agricultural University, Thrissur, 170 p

Bakare RI, Magbagbeola OA, Akinwande AI, Okunowo OW (2010) Nutritional and chemical evaluation of *Momordica charantia*. J Med Plant Res 4 (21):2189–2193

Behera TK, John JK, Simon, JE, Staub PW (2010) Bitter gourd: botany, horticulture, and breeding. In: Janick J (ed) Horticultural reviews, vol 37

Behera TK, Dey SS, Munshi AD, Gaikwad AB, Pal A, Singh I (2009) Sex inheritance and development of gynoecious hybrids in bitter gourd (*Momordica charantia* L.). Sci Hort 120:130–133

Behera TK, Gaikwad AB, Singh AK, Staub JE (2008) Comparative analysis of genetic diversity in Indian bitter gourd (*Momordica charantia* L.) using RAPD and ISSR markers for developing crop improvement strategies. Sci Hort 115:209–217

Bharathi LK, John KJ (2013) Momordica genus in Asia: an overview. Springer, New York, p 147

Bharathi LK, Munshi AD, Behera TK, Vinod John JK, Bhat KV, Das AB, Sidhu AS (2012) Production and preliminary characterization of novel inter-specific hybrids derived from *Momordica* species. Curr Sci 103:178–186

Bharathi LK, Vinod Munshi AD, Behera TK, Shanti C, Kattukunnel JJ, Das AB (2010) Cytomorphological evidence for segmental allopolyploid origin of teasel gourd (*Momordica subangulata* subsp. *renigera*). Euphytica 176:79–85

Cervancia CR, Barile GE (1993) Foraging behavior of *Trigona biroi* Friese (Apidae: Hymenoptera). Pollination in the tropics. Proceedings of the international symposium on pollination in tropics. IUSSI (India), p 78–80

Chakravarty HL (1959) Monograph on Indian Cucurbitaceae. In: Records of Botanical Survey of India, pp 86–99

Chakravarthy HL (1982) Cucurbitaceae: fascicles of flora of India 2. Botanical Survey of India, Howrah, p 94

Chakravarty HL (1990) Cucurbits of India and their role in the development of vegetable crops. In: Bates DM, Robinson RW, Jeffrey C (eds) Biology and utilization of Cucurbitaceae. Cornell University Press, Ithaca, New York, pp 325–334

Chang YM, Liou PC, Hsiao CH, ShE CT (1999) Observation of fruit anatomy nd development of bitter gourd. I. Fruit anatomy and fertilization of bitter gourd. J Agric Res China 48:23–31

Cronquist A (1988) The evolution and classification of flowering plants. The New York Botanical Garden, New York

Decker-Walters DS (1999) Cucurbits, sanskrit, and the Indo-Aryans. Econ Bot 53:98–112

Degner O (1947) Flora hawaiiensis. Book 5, Privately published, Honolulu, 185 p

Desai UT, Musmade AM (1998) Pumpkins, squashes and gourds. In: Salunkhe DK, Kadam SS (eds) Handbook of vegetable science and technology: production, composition, storage and processing. Marcel Dekker, New York, pp 273–298

Deshpande AA, Venkatasubbaiah K, Bankapur VM, Nalawadi UG (1979) Studies on floral biology of bitter gourd (*Momordica charantia* L.). Mys J Agri Sci 13:156–159

De Wilde WJJO, Duyfjes BEE (2002) Synopsis of *Momordica* (Cucurbitaceae) in SE-Asia and Malaysia. Bot Zhuro 87:132–148

Dey SS, Behera TK, Kaur C (2006) Genetic variability in ascorbic acid and carotenoids content in Indian bitter gourd (*Momordica charantia* L.) germplasm. Cucurbit Genet Coop Rep 28:91–93

Dey SS, Behera TK, Munshi AD (2010) Gynoecious inbred with better combining ability improves yield and earliness in bitter gourd (*Momordica charantia* L.). Euphytica 173:37–47

Dey SS, Behera TK, Munshi AD, Sirohi PS (2007) Studies on genetic divergence in bitter gourd (*Momordica charantia* L.). Indian J Hort 64:53–57

Deyto CR, Cervancia CR (2009) Floral biology and pollination of Ampalaya (*Momordica charantia* L.). Philipp Agri Sci 92(1): 8–18

Durry H (1864) Handbook of Indian flora, vol 1. Bishen Singh and Mahendrapal Singh Publishers, New Delhi, p 181

Faegri K, van der Pijil L (1979) The principles of pollination ecology 3rd edn. Pergamon Press, Great Britain, 244 p

Fonseka HH, Fonseka RM (2011) Studies on deterioration and germination of bitter gourd seeds (*Momordica charantia* L.) during storage. Acta Hort 98:31–38

Forbes MF, Cervancia CR (1994) Foraging behavior of *Apis cerana* F. and *Apis mellifera* L. (Apidae: Hymenoptera) in Majayjay, Laguna. Phil J Sci 123 (1):21–27

Goulson D (1999) Foraging strategies of insects for gathering nectar and pollen, and implications for plant ecology and evolution. Perspect Plant Ecol Evol Systemat 2(2):185–209

Grover JK, Yadav SP (2004) Pharmacological actions and potential uses of *Momordica charantia*: a review. J Ethnopharmacol 93:123–132

Gill LS, Karatela YY (1982) Epidermis structure and stomatal ontogeny in some Nigerian Cucurbitaceae. Willdenowia 12(2):303–310

Giuliani C, Tani C, Maleci Bini L (2016) Micromorphology and anatomy of fruits and seeds of bitter melon (*Momordica charantia* L., Cucurbitaceae). Acta Soc Bot Pol 85(1):3490

Gopalan C, Sastri BVS, Balasubramanian SC (1982) Nutritive values of Indian foods. National Institute of Nutrition (ICMA), Hyderabad, AP, India, p 161p

Hsu CC, Chen CL, Chen JJ, Sung MJ (2003) Accelerated aging-enhanced lipid peroxidation in bitter gourd seeds and effects of priming and hot water soaking treatments. Sci Hort 98:201–212

https://portal.wiktrop.org/species/show/216. Accessed on 22 April 2019

https://uses.plantnet-project.org/en/Momordica_charantia_(PROTA). Accessed on 3 May 2019

Husain J, Tickle IJ, Wood SP (1994) Crystal structure of momordin, a type I ribosome inactivating protein from the seeds of *Momordica charantia*. FEBS Lett 342:154–158

Islam MR, Hossain MS, Buhiyan MSR, Husna A, Syed MA (2009) Genetic variability and path coefficient analysis of bitter gourd (*Momordica charantia* L.). Intl J Sustain Agri 1:53–57

Islam S, Mis MAB, Das MR, Hossain T, Ahmed JU, Hossain MM (2014) Sex phenology of bitter gourd (*Momordica charantia* L.) landraces and its relation to yield potential and fruit quality. Pak J Agri Sci 51 (3):651–658

Iwamoto E, Ishida T (2006) Development of gynoecious inbred line in balsam pear (*Momordica charantia* L.). Hort Res 5:101–104

Jeffrey C (1980) A review of the Cucurbitaceae. J Linn Soc Bot 81:233–237

Jeffrey C, De Wilde WIJO (2006) A review of the subtribe Thalidianthinae (Cucurbitaceae). Bot Z 91:766–776

Joseph JK (2005) Studies on ecogeography and genetic diversity of the genus *Momordica* L. in India. Dissertation, Mahatma Gandhi University, Kottayam, Kerala, 312 p

Joseph JK, Antony VT (2010) A taxonomic revision of the genus *Momordica* L. (Cucurbitaceae) in India. Indian J Plant Genet Resour 23(2):172–184

Joseph JK, Antony VT (2008) Ethnobotanical investigations in the genus *Momordica* L. in the Southern Western Ghats of India. Genet Res Crop Evol 55 (5):713–721

Katiyar D, Singh V, Ali M (2017) Phytochemical and pharmacological profile of *Momordica charantia*: a review. biochemistry and therapeutic uses of medicinal plants, 1–34

Kearns CW, Inouye DW (1993) Techniques for pollination biologists. University Press of Colorado, Niwot, Colorado, 583 p

Keeding GB, Krawinkel MB (2006) Bitter gourd (*Momordica charantia*): a dietary approach to hyperglycemia. Nutr Rev 64:331–337

Khan MR, Omoloso AD (1998) *Momordica charantia* and *Allium sativum*: broad spectrum antibacterial activity. Kor J Pharmacol 29(3):155–158

Klomann SD, Mueller AS, Pallauf J, Krawinkel MB (2010) Antidiabetic effects of bitter gourd extracts in insulin resistant db/db mice. Br J Nutr 10(4):1613–1620

Krishnendu JR, Nandini PV (2016) Nutritional composition of bitter gourd types (*Momordica charantia* L.). Intl J Adv Eng Res Sci 3:95–104

Kumar N, Pandey AK (2002) Genus *Momordica* in India: diversity and conservation. In: Das AP (ed) Perspectives of plant biodiversity. Bishen Singh Mahendra Pal Singh Publishers, Dehradun, pp 35–43

Larkcom J (1991) Oriental vegetables: the complete guide for garden and kitchen. John Murray, London

Lin JM, Sung JM (2001) Pre-sowing treatments for improving emergence of bitter gourd seedlings under optimal and sub-optimal temperatures. Seed Sci Technol 29:39–50

Linnaeus C (1753) Species planetarium (vol 2). Stockholm, Facsimile of the first edn Published by Ray Society, London, in 1959

Marr KL, Xia YM, Bhattarai NK (2004) Allozyme, morphological and nutritional analysis bearing on the domestication of *Momordica charantia* L. (Cucurbitaceae). Leon Bot 58:435–455

Metcalfe CR, Chalk L (1950) Anatomy of dicotyledons 1. Clarendon Press, Oxford. https://archive.org/details/anatomyofthedico033552mbp

Mia MAB, Shamsuddin ZH (2011) Physio-morphological appraisal of aromatic fine rice (*Oryza sativa* L.) in relation to yield potential. Intl J Bot 7:223–229

Mia MAB, Das MR, Kamruzzaman M, Talukder NM (2012) Biochemical traits and physio-chemical attributes of local aromatic fine rice landraces in relation to yield potential. Amer J Plant Sci 3(12):1788–1795

Miniraj N, Prasanna KP, Peter KV (1993) Bitter gourd *Momordica* spp. In: Kalloo C, Bergh BO (eds) Genetic Improvement of vegetable plants. Pergamon Press, Oxford, UK, pp 239–246

Morton JF (1967) The balsam pear—an edible, medicinal and toxic plant. Econ Bot 21:57–68

Nath P, Soni SL, Charan R (1972) Effect of plant growth regulators, light and their interactions on seed germination in bitter gourd (*Momordica charantia* L.) In: Proceedings of 3rd symposium on subtropical horticulture, vol III. Indian Institute of Horticultural Research, Bangalore, pp 173–182

NBPGR (2001) Minimal descriptors of agri-horticultural crops, part-II. vegetable crops. National Bureau of Plant Genetic Resources, Pusa, New Delhi, p 262

Njoroge GN, Luijk MN (2004) *Momordica charantia* L. In: Grubben GJH, Denton OA (eds) PROTA 2: vegetables/legumes. Wageningen, PROTA, pp 131–135

Oliver AEP (1960) Medicinal plants in Nigeria. College of Arts and Science Technology, River State, Nigeria, p 85

Palada MC, Chang LC (2003) Suggested cultural practices for bitter gourd. Taiwan: Asian Vegetable Research and Development Center (AVRDC) 547 (3): 1–5

Parkash A, Ng TB, Tso WW (2002) Purification and characterization of charantin, a napin-like ribosome-inactivating peptide from bitter gourd (*Momordica charantia*) seeds. J Peptide Res 59:197–202

Peter KY, Sadhu MK, Raj M, Prasanna KP (1998) Improvement and cultivation: bittergourd, snake gourd, pointed gourd and ivy gourd. In: Nayarand NM, More TA (eds) Cucurbits. Oxford and IBH Publishing House, New Delhi, pp 187–195

Pillai OAA, Irulappan I, Jayapal R (1978) Studies on the floral biology of bitter gourd (*Momordica charantia* L.) varieties. Madras Agri J 65:168–171

Poyrozki E, Derdovski C (2017) Morpho-anatomical investigations on *Momordica charantia* L. (Cucurbitaceae). Anadolu Univ J Sci Technol Life Sci Biotechnol 5(1):23–30

Raj NM, Prasanna KP, Peter KV (1993) Bitter gourd. In: Kalloo G, Bergh BO (eds) Genetic improvement of vegetable plants. Pergamon Press, Oxford, pp 239–246

Rajput JC, Paranjape SP, Jamadagni BM (1996) Variability, heritability and scope of improvement for yield components in bitter gourd (*Momordica charantia* L.). Ann Agri Res 17:90–93

Ram D, Kumar S, Banerjee MK, Kalloo G (2000) Occurance, identification and preliminary characterization of gynoecium in bitter gourd (*Momordica charantia* L.). Intl J Agri Sci 72:348–349

Ram D, Kumar S, Banerjee MK, Singh B, Singh S (2002) Developing bitter gourd (*Momordica charantia* L.) populations with very high proportion of pistillate flowers. Cucurbit Genet Coop Rep 25:65–66

Raman A, Lau C (1996) Anti-diabetic properties and phytochemistry of *Momordica charantia* L. (Cucurbitaceae). Phytomedicine 2:329–362

Rasco AO, Castillo PS (1990) Flowering patterns and vine pruning effects in bitter gourd (*Momordico charantia* L.) varieties 'Sta. Rita' and 'MaUling'. Philip Agri 73:3–4

Rasul MG, Hiramatsu M, Okubo H (2004) Morphological and physiological variation in Kakrol (*Momordica diocia* Roxb.). J Fac Agri 49:1–11

Renner SS, Pandey AK (2013) The Cucurbitaceae of India: accepted names, synonyms, geographic distribution and information on images and DNA sequences. Phytokeys 20:53–118

Reyes ME, Gildemacher CBH, Jansen GJ (1994) *Momordica* L. In: Siemonsma JS, Piluek K (eds) Plant resources of South-East Asia: vegetables. Pudoc Scientific Publishers, Wageningon, Netherlands, pp 206–210

Rheede VHA (1688) Hortus malabaricus, vol 8. Joannis VS, Joannis DV (eds). Amsterdam, pp 17–36

Robinson RW, Decker-Walters DS (1999) Cucurbits. CABI Publishers, Wallingford, p 101

Sahib NG, Saari N, Ismail A, Khatib A, Mahomoodally, Hamid AA (2012) Plant metabolites as potential antiobesity agents. Sci World J 1–8. http://dx.doi.org/10.1100/2012/436039

Sands WN (1928) The bitter cucumber of Paris. Malayan Agric. J. 16:32–39

Schaefer H, Renner SS (2010) A three-genome phylogeny of *Momordica* (Cucurbitaceae) suggests seven returns from dioecy to monoecy and recent long-distance dispersal to Asia. Mol Phylogenet Evol 54:553–560

Schaefer H, Renner SS (2011) Phylogenetic relationships in the order Cucurbitales and a new classification of the gourd family (Cucurbitaceae). Taxon 60(1):122–138

Schaefer H, Heibel C, Renner SS (2009) Gourds afloat: a dated phylogeny reveals an Asian origin of the gourd family (Cucurbitaceae) and numerous oversea dispersal events. Proc Rev Soc 276:843–851

Shawaf IIS, Baker IR (1981) Combining ability and genetic variances of GXH F1 hybrids for parthenocarpic yield in gynoecious pickling cucumber for once

over mechanical harvest. J Amer Soc Hort Sci 106:365–390

Shiefriss O, Galun E (1956) Sex expression in the cucumber. J Amer Soc Hort Sci 67:479–486

Sidhu GK, Pathak M (2016) Genetic diversity analysis in bitter gourd (*Momordica charantia* L.) using morphological traits. Intl J Agri Inn Res 41:59–63

Singh SK, Ram HH (2005) Seed quality attributes in bitter gourd (*Momordica charantia* L.). Seed Res 33:92–95

Soyimchiten L, Pradheep K, John JK, Nayar ER (2015) An occurrence of Indo-Chinese taxon *Momordica subangulata* Blume subsp. *subangulata* (Cucurbitaceae) in Nagaland: a new distribution record from India. J Threat Taxa 7(14):8182–8184

Srivatsava VK, Srivatsava LS (1976) Genetic parameters, correlation coefficient and path coefficient analysis in bitter gourd (*Momordica charantia* L.). Indian J Hort 33(1):66–70

Subhakar G, Sreedevi K, Manjula K, Reddy NPE (2011) Pollinator diversity and abundance in bitter gourd, Momordica charantia Linn. Pest Manag Horticul Ecosyst 17(1):23–27

Temitope AG, Sheriff OL, Azeezat YF, Taofiks A, Fatimah AI (2013) Cardioprotective properties of *Momordica charantia* in albino rats. Afr J Sci Res 11(1):600–610

Tewari GN, Singh K (1983) Role of pollinators in vegetable seed production. Indian Bee J 45:51

Walters TW, Decker-Walters DS (1988) Balsam pear (*Momordica charantia*, Cucurbitaceae). Econ Bot 42:286–288

Wu AJ, Ng LT (2008) Antioxidant and free radical scavenging activities of wild bitter gourd (*Momordica charantia* var. *abbreviate* Ser.) in Taiwan. LWT—Food Sci Technol 41:323–330

Xie H, Huang S, Deng H, Wu Z, Ji A (1998) Study on chemical components of *Momordica charantia*. Zhong Yao Cai 21:458–459

Yang SL, Walters WL (1992) Ethnobotany and the economic role of the Cucurbitaceae of China. Econ Bot 46:349–367

Yasuda A (1999) On the comparative anatomy of the Cucurbitaceae, wild and cultivated, in Japan. J Coll Sci Imper Univ 18:1–56

Yuan YR, He YN, Xiong JP, Xia ZX (1999) Three-dimensional structure of betamomorcharin at 2.55 A resolution. Acta Crystallographica Sec D-Biol Crystallo 55:1144–1151

Zhou WB, Lou S, Luo JN (1998) An early maturing and high yielding bitter gourd hybrid Cuilli No 1. Plant Breed Abstr 68:1002–1005

Medicinal Properties of Bitter Gourd: Bioactives and Their Actions

3

Vidhu Aeri and Richa Raj

Abstract

Bitter melon (*Momordica charantia* L., Family: Cucurbitaceae) is traditionally used as a medicinal food in different systems of medicine. It has significant importance in providing basic nutrients and prevention of various ailments. It contains a variety of bioactive compounds including alkaloids, polypeptides, vitamins, and minerals. A diversity of bioactive compounds comprises two classes of saponins: Oleanane and Cucurbitane-type triterpenoids. The present chapter shall draw a link of the bioactive compounds to its pharmacological effects like antidiabetic, anticancer, antiviral, anti-inflammatory, analgesic, hypolipidemic, and hypocholesterolemic effects, and an insight to understand the mechanism of action.

3.1 Introduction

Momordica charantia L., (MC) a tropical vegetable, is known by different names in traditional systems of medicine across the globe: bitter melon and balsam pear (English), Karella (Hindi or Urdu), Nigauri or Goya (Japanese), Ku gua (Mandarin), Ko guai (Taiwanese), Kho qua (Vietnamese), Ampalaya (Filipino), and Assorossie in French (Grover et al. 2004).

In addition to its major use as an antidiabetic agent, bitter gourd has been used traditionally as a treatment for tumors, asthma, skin infections, gastrointestinal problems, and hypertension (Kumar and Bhowmik 2010). Fruits are generally used in Asia; in addition leaves, fruits and roots are also used to treat fevers (Raman and Lau 1996).

According to Ayurvedic system of medicine, roots are useful in treatment of eye relatdiseases. The fruit is bitter, cooling, digestible, laxative, antipyretic, anthelmintic, appetizer, cures biliousness, blood diseases, anemia, urinary discharges, asthma, ulcers, and bronchitis. In Unani system of medicine, fruits are carminative, tonic, stomachic, aphrodisiac, anthelmintic, astringent to bowels, and useful in treatment of syphilis, rheumatism, and spleen troubles (Gupta et al. 2011).

Bitter gourd contains a range of biologically active plant constituents which includes: Oleanane and cucurbitane-type triterpenoids, proteins, steroids, alkaloids, saponins, flavonoids, and acids.

Bitter gourd has been extensively investigated for its variety of health and pharmacological properties: anticancer (Biswas et al. 1991; Singh et al. 1998; Choi et al. 2002; Nagasawa et al.

V. Aeri (✉) · R. Raj
School of Pharmaceutical Education and Research, Jamia Hamdard, New Delhi, India
e-mail: vaeri@jamiahamdard.ac.in

R. Raj
e-mail: richa14.raj92@gmail.com

© Springer Nature Switzerland AG 2020
C. Kole et al. (eds.), *The Bitter Gourd Genome*, Compendium of Plant Genomes,
https://doi.org/10.1007/978-3-030-15062-4_3

2002; Pongnikorn et al. 2003; Deep et al. 2004; Konishi et al. 2004; Limtrakul et al. 2004; Senanayake et al. 2004; Chen and Li 2005; Yasui et al. 2005; Fernandes et al. 2007; Gadang et al. 2011), antiviral (Grover and Yadav 2004; Beloin et al. 2005), anti-inflammatory and analgesic (Kobori et al. 2008; Lin and Tang 2008), and hypolipidemic and hypocholesterolemic effects (Nerurkar et al. 2010).

Furthermore, numerous studies have been conducted on the effectiveness of fresh, juiced, or dried bitter gourd in diabetic animals and in type 2 diabetic human subjects (Welihinda and Karunanayake 1986; Ali et al. 1993; Chen et al. 2003; Virdi et al. 2003; Chaturvedi et al. 2004; Grover and Yadav 2004; Sathishsekar and Subramanian 2005; Harinantenaina et al. 2006; Dans and Villarruz 2007).

Scientists are investigating the underlying mechanism of the action of bioactive compounds. Pure protein p-insulin, steroidal saponins, and alkaloids exhibit blood sugar lowering activity by inhibiting the activity of glucosidase, inhibits the absorption of glucose.

3.2   Traditional and Ethnomedicinal Uses

Traditionally, the juice of bitter gourd leaves is used for applying on the skin for treating insect bites, bee stings, burns, contact rashes, and wounds (Aljohi et al. 2016). Decoction of its leaves and fruits is drunk as preventative or treatment of stomachache, toothache, liver diseases, diabetes, hypertension, and cancer (Ahmad et al. 2016). In Asia, bitter gourd has been considered effective for the management and prevention of malaria (Aljohi et al. 2016). In Colombia and Panama, tea from bitter melon leaves has been considered to be useful for the treatment of malaria. Crude ethanolic extracts from Momordica has exhibited strong antimalarial activity. Oral administration of leaves showed strong antimalarial activity reduces the levels of parasitemia in plasmodium-infected mice (Olasehinde et al. 2014).

Bitter gourd has been used in various Asian traditional medicine systems for a long time, for preventing and treating various diseases. Its fruits and pulp are used in treating asthma, constipation, colic, diabetes, cough, fever (malaria), gout, helminthiases, leprosy, inflammation, skin diseases, ulcer, and wound. It has also shown hypoglycaemic (antidiabetic) properties in animal as well as human studies. Juice of bitter gourd is used to treat piles, and as a blood purifier due to its bitter tonic properties, also beneficial in treating and preventing the liver damage.

In India, bitter gourd is used by tribal people for abortions, birth control, increasing milk flow, vaginal discharge, menstrual disorders, constipation, food, hyperglycemia, diabetes, jaundice, stones, kidney, liver, fever (malaria), eczema, gout, fat loss, hemorrhoids, hydrophobia, intestinal parasites, skin, pneumonia, leprosy, psoriasis, rheumatism, scabies, piles, snakebite, and as an antihelminthic (Rawlings et al. 2016). In Sri Lanka, it is used as a tonic, emetic, and laxative.

In Guyana traditional medicine, leaf tea is used for diabetes, to expel intestinal gas, and to promote menstruation. It is used topically for sores, fungal and wound infections (Prabhakar et al. 2008).

Bitter gourd seeds have shown its effectiveness in treatment of ulcers, liver and spleen problems, diabetes, high cholesterol, intestinal parasites, heal wounds, and stomachache.

Roots of bitter gourd are also used in the treatment of syphilis, rheumatism, ulcer, boils, and septic swellings. Bitter gourd juice helps to reduce the problem of pyorrhea (bleeding from the gums). Bitter gourd capsules and tinctures are widely available in the USA for the treatment of diabetes, cold, flu, viruses, tumors, cancer, high cholesterol, and psoriasis.

3.3   Bioactive Compounds

Each and every part of the plant indicates potential health benefits. It might be beneficial to include bitter gourd in our daily life (Chaturvedi. 2009; Braca et al. 2008; Alam et al. 2015). It contains vitamins, minerals, and flavonoids (Aziz and Karboune 2016). It is a good source of phenolic compounds, particularly gallic acid (Puri et al. 2011). It also contains a variety of other bioactive compounds, especially saponins, peptides, and alkaloids (Cao et al. 2014).

A diversity of bioactive compounds is present in bitter gourd, which mainly comprise two classes of saponins: Oleanane and cucurbitane-type triterpenoids.

3.3.1 Cucurbitane Triterpenoids

Cucurbitacins are most common in the species of Bryonia, Citrullus, Cucumis, Cucurbita, Luffa, Echinocystis, and Lagenaria of the family Cucurbitaceae. The plants of the genus Momordica contain a special group of cucurbitacins called momordicosides. The level of cucurbitacins varies between tissues and may be concentrated in fruits and roots of mature plants. In fruits, their highest concentration is achieved on maturity. Seeds generally contain very low concentration of cucurbitacins. Cucurbitacin producing plants have also been identified in the members of the family Scrophulariaceae, Begoniaceae, Primulaceae, Liliaceae, Tropaeolaceae, and Rosaceae. The seeds of certain cruciferous plants, like Iberis species and Lepidium sativum also contain cucurbitacins (Kee and Hongtao 2008). The cucurbitacins are formed in situ and are not transported to other parts of the plant (Frohne 1983).

In general, cucurbitacins contain a basic 19-$(10 \rightarrow 9\beta)$–abeo–10α–lanost–5–ene ring skeleton. A common feature among all compounds in the category of cucurbitacins is the presence of a double bond (Frohne 1983; Zhu and Zhang 2013; Dinan et al. 2001). The cucurbitacins differ from most of the other tetracyclic triterpenes by being highly unsaturated and contain numerous keto–, hydroxyl–, and acetoxy–groups (Jorn et al. 2006). A special group of cucurbitacins is called as momordicosides, named after their occurrence in Momordica charantia. Momordicosides have never been identified in any other plant species. The common feature of

momordicosides is that C_{19} has been oxidized to an aldehyde group.

Various biological activities like anti-inflammatory (Jayaprakasam et al. 2003; Park et al. 2004; Yuan et al. 2006; Escandell et al. 2008; Ri'os et al. 1990), antitumor (Rehm 1957; Bowman et al. 1999; Duncan et al. 1996; Bowman et al. 2000; Liu et al. 2000; Turkson and Jove 2000; Higashio 2002; Blaskovich et al. 2003; Sun et al. 2005; Dong et al. 2010; Duangmano and Dakeng 2010), antineurogenrative activity (Faden and Ioane 2015) and anti-artherosclerotic activity (Esterbauer 1993; Tannin-Spitz et al. 2007; Saba and Oridupa 2010) have been documented.

3.3.2 Antidiabetic Activity Attributed to Cucurbitacins

There have been a plethora of reports on the role of cucurbitacins for their cytotoxic, hepatoprotective, cardiovascular, and antidiabetic effects (Park et al. 2004). Cucurbitane triterpenoids present in Momordica fruits are noted for antidiabetic and anticancer activities, this may provide leads as a class of therapeutics for diabetes and obesity (Zhu et al. 1990; Harinantenaina et al. 2006; Jianchao et al. 2008; Tan et al. 2008). The 5'-adenosine monophosphate-activated protein kinase (AMPK) pathway is suggested as a probable mechanism for the stimulation of GLUT4 translocation by triterpenoids from M. charantia. It is particularly interesting in relation to diabetes and obesity because activation of AMPK increases fatty acid oxidation, inhibits lipid synthesis, and can improve insulin action (Iglesias and Ye 2002; Ye et al. 2005).

3.3.3 Oleanane Triterpenoids

It is well known that the naturally occurring modified triterpenes in plants have a wide diversity of chemical structures and biological functions. The lupane-, oleanane-, and ursane-type triterpenes are the three major members of natural triterpenes with a wide range of biological properties. Some oleanane-type compounds are of interest due to their antitumor, anti-HIV, anti-inflammation, antioxidation, antihyperglycemia, and cardiovascular properties.

> The chemistry of oleanane- and ursane-type triterpenoids has been actively explored in recent years, and biological and pharmacological activities of these compounds have been found to span a variety of properties. These include antitumor, antiviral, anti-inflammatory, hepatoprotective, gastroprotective, antimicrobial, antidiabetic, and hemolytic properties as well as many others.

3.4 Major Bioactive Potential Candidates and Their Mechanism of Action

Different plant parts and respective extracts exhibited important biological activities: leaf extract exhibited antimalarial action against *Plasmodium falciparum*, juice was found to be effective in wound healing, psoriasis, scabies, and ringworm, and also extract was found to be anti-inflammatory by inhibiting activation of nuclear transcription factor-κB (NF-κB) through stimulation by tumor necrosis factor-α (TNFα) (Martínez-Abundis et al. 2016).

Reports indicated antiprotozoal activity of whole extract, methanol, water, and ethanol extracts of the bitter gourd leaves exhibited antibacterial action against *Salmonella, pseudomonas aeruginosa, E. coli, Bacillus, and Streptococcus chain* (Brandao et al. 2016).

Dried bitter gourd powder with honey showed substantial anti-ulcerogenic activity (Alam et al. 2009). High intake up to 500 mg/kg of the extract showed antiulcer activity and significantly reduced total acidity and pepsin activity, while lower dose did not inhibit the ulcer proliferation.

3.4.1 Antidiabetic Activity and Antihyperlipedimic Activity

Pure protein p-insulin significantly lowers blood glucose level, similar to insulin by inhibiting the activity of glucosidase which inhibits the absorption of glucose. Steroidal saponins (charantin) effectively control sugar level in blood, act like peptides, alkaloids effectively control sugar level in blood.

Saponins: oleanane and cucurbitane-type triterpenoids are present in whole plant. one new cucurbitane-type triterpenoid glycoside, momordicoside U (Namsa et al. 2011), along with five well-known cucurbitane-type triterpenoids and associated glycosides, kuguaglycoside G (Tambor et al. 2016), 3- hydroxycucurbita-5,24-dien-19-al-7,23-di-O-β-glucopyranoside (Aziz and Karboune 2016), momordicine (Sabourian et al. 2016), 3β,7 β,25-trihydroxycucurbita-5,23 (E)-dien-19-al [2], and momordicine II (Pandit et al. 2016) have been reported and indicate marked antidiabetic effect.

Three new cucurbitane triterpenoids and one new steroidal glycoside were isolated together with ten known compounds from bitter gourd (Liu et al. 2009; kamath et al. 2008; Li et al. 2016).

Hyperlipidemia is associated with diabetes. High blood lipid concentration leads to atherosclerosis, cerebrovascular and ischemic heart diseases. Different fractions of *M. charantia* significantly showed antihyperlipidemic effect: metformin, flavonoids, saponins, tannins, triterpenes, and alkaloids effect total cholesterol level in diabetic rats.

The probable mechanism of action suggests the repair of damaged β-cells; thereby, increasing the levels of insulin and its sensitivity (Chaturvedi 2012). Bitter gourd also stimulates the release and synthesis of adiponectin and thyroid hormones. Bitter gourd enhances the action of adenosine-5-monophosphate kinase (AMPK) that is associated with fat release from fatty tissues and glucose uptake and thus causing in weight loss (Yang et al. 2015).

Hepatic production of triglycerides also contributes to the hyperlipidemic effect of HIV-1-protease inhibitors and that contain lipoprotein instead of lipoprotein clearance (Bai et al. 2016). The decrease in cholesterol and triglycerides levels was mediated through enhanced excretion of fecal lipid excretion and their lymphatic transport (Sato et al. 2011; Chao et al. 2014).

In HepG2 cells, the bioactive constituents decrease apolipoprotein C-III and decrease liver secretion of apolipoprotein B (Apo-B). Apo-B protein known as lipoprotein is used for the production of LDL. Apo-C-III is a lipoprotein which is involved in the synthesis of LDL and found to be present in VLDL. The bioactive constituents increase Apo lipoprotein A-1 (Apo A-1) which is basic protein component compulsory for HDL synthesis (Dar et al. 2014).

3.4.2 Antioxidant and Anti-inflammatory Activity

Total phenolic compounds like gallic acid, epicatechin, chlorogenic acid, catechin, and gentisic acid are free radical scavenging agents, present in whole fruit, seeds, and leaves, and regulates impaired antioxidant status and suppress fat accumulation. Flavonoids regulate blood cholesterol; thereby, providing protection from cardiovascular disorders like atherosclerosis (Gil et al. 2002).

Leaf, fruit, seed, and ethanolic extracts have been reported to possess antioxidant and anti-inflammatory activity because of the presence of phenolic compounds (Qader et al. 2011; Chunthorng-Orn et al. 2012; Aljohi et al. 2016). Other solvent extracts are endorsed to the existence of higher amounts of flavonoids and phenolics.

The levels of TBARS, hydroperoxides, ALT, AST, and GPx (responsible for liver damage and lipid peroxidation) are normalized on oral administration of bitter gourd (Thenmozhi and Subramanian 2012; Sagor et al. 2015).

The highest value based on DPPH radicals-scavenging activity and ferric-reducing power was observed for leaf extract, while the green fruit extract showed the highest antioxidant activity based on hydroxyl radical-scavenging activity, β-carotene-linoleate bleaching assay, and total antioxidant capacity (Yadav et al. 2016)

Aqueous and ethanolic extract possess significant DPPH radical-scavenging activity and iron chelating as compared to Vit. E (Kamal et al. 2011; Yehye et al. 2016).

3.4.3 Anticancer Activity

Seed oil–germacrene, trans-nerolidol apiole, and cis-dihydrocarveol (Mesia et al. 2008; Ahmad et al. 2012) and alpha-eleostearic acid are abundant in oil, which has strong properties of lowering blood fat, inhibiting the proliferation of tumor cell, inhibiting CVD, and have anticancer and anti-inflammatory properties (Liu et al. 2010).

Stronger antioxygenic activity has been shown by bitter gourd seed powder and pulp. Ethanol and aqueous extracts showed great potential as natural antioxidants to inhibit lipid peroxidation in foods (Padmashree et al. 2011; Shodehinde et al. 2016).

Momordin I, α, β momorcharin and cucurbitacin B were found to be active against breast cancer, indicating antiproliferative action; thereby, inhibiting the growth of breast cancer by encouraging autophagic cell death.

Kuguacin J: has ability to constrain the prostate cancer growth. Seed, pericarp, and placenta extracts: induce apoptosis in HL60 human leukemia cells. In HL60 cells, apoptosis is induced by α-eleostearic acid.

Polysaccharide (MCP2): antitumor activity. The growth of Hela cells and HepG2 cells has been inhibited by MCP2 and its sulphated drivatives, which indicated that the anti-tumour activity of MCP2 might be enhanced by sulphated modification.

MAP-30: Anticancer, inhibition of expression of the HER2 gene and inhibition of cancer cell proliferation in vitro, MCP30 may restore normal PTEN signaling as demonstrated by decreased activity of Akt by dephosphorylation at Ser-473, increased Ser-9 phosphorylation of GSK-3b, inhibition of canonical Wnt signaling, and decreased expression of Cyclin-D1 and c-Myc in the neoplastic prostate cells has been tested highly meta-static human breast tumor MDA-MB -231 cells and estrogen-independent cells.

The exposure of HepG2.2.15 cells to MAP30 resulted in inhibition of HBV DNA replication and HBsAg secretion. Lower dose of MAP30 (8.0 microg/ml) could inhibit the expression of HBsAg and HBeAg. (Fan et al. 2009; Liu et al. 2016) Previous studies have shown that extracts of wild bitter gourd suppress lymphocyte proliferation, and macrophage and lymphocyte activity (Chao et al. 2011; Bhattacharya et al. 2016). Effects of MCP30 on HDAC1 in prostate-derived cell lines were observed because this particular HDAC was previously shown to be overexpressed in human premalignant and malignant prostate lesions (Qader et al. 2011, Huang et al. 2016) with the highest increase in expression in hormone-refractory prostate cancer (Kavitha et al. 2011; Tabackman et al. 2016).

3.4.4 Anti-inflammatory Activity

Inhibition of the overproduction of inflammatory mediators, especially pro-inflammatory cytokines IL-1b, IL-6, and TNF-a, may prevent or suppress a variety of inflammatory diseases (Chao et al. 2014). Previous reports indicated that the extract of bitter melon plant inhibits activation of nuclear transcription factor-κB (NF-κB) through stimulated by tumor necrosis factor-α (TNFα).

The ethanolic extracts (250 and 500 mg/kg) showed an analgesic and antipyretic effect (Nhiem et al. 2012). Bitter gourd is beneficial for reducing LPS-induced inflammatory responses by modulating NF-kappaB activation (Chao et al. 2014). Placenta extract of bitter gourd suppressed lipopolysaccharide (LPS)-induced TNFalpha production in RAW 264.7 macrophage-like cells. Butanol-soluble fraction of bitter gourd placenta extract strongly suppresses LPS-induced TNFalpha production in RAW 264.7 cells. Gene expression analysis using a fibrous DNA microarray showed that the bitter gourd butanol

fraction suppressed expression of various LPS-induced inflammatory genes, such as those for TNF, IL1alpha, IL1beta, G1p2, and Ccl5.

3.4.5 Antiviral and Antimicrobial Activity

Five cucurbitacins, kuguacins A-E (1-5), together with three known analogs, 3beta, 7beta, 25-trihydroxycucurbita-5, (23E)-diene-19- al, [6] 3beta, 25-dihydroxy-5beta, 19-epoxycucurbita-6, (23E)-diene, and momordicine I, were isolated from roots of bitter gourd (Kai et al. 2011; Laverdure et al. 2016).

Momordicoside A and B–antiviral against various viruses infections including Epstein-Barr, herpes, and HIV viruses, increasing interferon production and natural killer cell activity.

α-momorcharin: antiviral: effective in inhibiting the fungal and bacterial growth. Their potential against *Fusarium solani* (IC50 value: 6.23 μM), *Fusarium oxysporum* (IC50 value: 4.15 μM), and *Pseudomonas aeruginosa* (IC50 value: 0.59 μM) is described due to its ribosome inactivating protein (RIP) ability. α-momorcharin due to its ribosome inactivating protein (RIP) ability is effectual in inhibiting the fungal and bacterial growth. Santos et al. (2012) further elaborated that bitter gourd is useful to treat fungal and parasitic diseases like Chagas disease, caused by *Trypanosoma cruzi*.

Gentisic acid: a biosynthetic derivative of salicylic acid produces pathogen inducible signals for the activation of plant defenses.

Bitter gourd also has cytotoxic and antiprotozoal activities. Bitter gourd exhibited antiprotozoal activity against *Trypanosoma brucei* (Phillips et al. 2013). The antimicrobial activity of the bitter gourd extract was tested on four clinical strains of *Cryptococcus neoformans, Klebsiella pneumoniae, Proteus vulgaris,* and *Salmonella typhi* and four reference microorganisms *Staphylococcus aureus, Pseudomonas aeruginosa, Candida albicans,* and *Escherichia coli* (Mwambete. 2009; Yaldiz et al. 2015).

The aqueous extract of bitter gourd and momordicatin is effective against leishmaniasis caused by *Leishmania donovani*. The mode of action includes inhibition of SOD which is one of the significant enzymes of the oxidative burst. Bitter gourd and its bioactive molecules are potential candidates as chemotherapeutics against leishmaniasis (Gupta et al. 2010; Dandawate et al. 2016).

The bitter gourd extract may act as a useful biolarvicide against mosquitoes. Investigations had been conducted against two types of mosquito vectors such as *Culex quinquefasciatus* and *Anophels stephensi* (Maurya et al. 2009).

Crude ethanolic extracts of bitter gourd have strong antimalarial activity. Oral administration of leaves reduced the levels of parasistemia in plasmodium-infected mice (Olasehinde et al. 2014). The IC50 value for *Momordica* against the plasmodium was 68.4 µg/mL (Gbedema et al. 2015; Dandawate et al. 2016).

Some compounds from bitter gourd showed moderate anti-HIV-1 activity with EC values of 8.45 and 25.62 microg/ml (Wang et al. 2010) and exerted minimal cytotoxicity (Chen et al. 2008; Meng et al. 2014).

Amongst the various ribosome inactivating proteins (RIPs) isolated from bitter gourd, MAP30 (*Momordica* protein of 30 kDa) displayed antitumor activity (Nerurkar et al. 2010; Fan et al. 2015).

Adult T-cell leukemia (ATL) is caused by human T-cell leukemia virus type I (HTLV-I) infection and is resistant to conventional chemotherapy. The bitter gourd seeds suppressed the proliferation of three cell lines. Hydroxy-pentanorcucurbit-5-en-3-one and dioxo-pentanorcucurbit-5-en-22-oic acid extracted from bitter gourd showed potent cytoprotective activity in tert-butyl hydroperoxide (t-BHP)-induced hepatotoxicity of HepG2 cells.

Anti-anthelmintic effect in gastrointestinal system is caused by nematodes, cestodes, and trematodes (Kappagoda et al. 2011; Rajeswari. 2014; Veerakumari 2015). Bitter gourd is considered as an important therapeutic medicinal food with antithelmintic action due to the presence of bioactive molecules like saponins, namely momordin, momordicoside, momordicin, kuguacin, karavilsodie, and karavilagenin. The mechanism of action being inhibition of arachidonic acid metabolism, mico nicotinic agonists, oxidative phosphorylation inhibition, increased calcium permeability, acetyl cholinesterase inhibitors, and β-tubulin binding with bioactive molecules (Melzig et al. 2001; Hrckova and Velebny 2013; Bauri et al. 2015).

Alkaloids including steroidal alkaloid and oligoglycosides have neurotoxic properties, act as acetylcholinesterase inhibitor. Alkaloids act as an antioxidant and are capable of decreasing the generation of nitrate which may interfere in homeostasis that is important for helminthes development (Wink 2012; Jain et al. 2013).

Flavonoid compounds including apigenin can inhibit larval growth and inhibit the arachidonic acid metabolism which may lead to the degeneration of neurons in the worm's body and lead to death (Ferrándiz and Alcaraz 1991; Yoon et al. 2006).

Tannins exhibit anthelmintic effect by reducing migratory ability and survival of newly hatched larvae and inhibit energy generation of worms by uncoupling the oxidation phosphorylation and bind to glycoprotein on the cuticles of the worms and lead to death (Iqbal et al. 2004; Williams et al. 2014).

3.4.6 Wound Healing

The juice of bitter gourd has a healing potential against psoriasis, scabies, and ringworm and is used in the inhibition of leprosy (Ilhan and Bolat 2015). Ethanolic extract of bitter gourd fruits and leaves is very effective in wound healing Hussan et al. (2014). Ribosome inactivating proteins inhibit the synthesis of proteins which promote viral diseases.

Compounds like MAP-30, MRK-29, momorcharin, and lectin have a protective effect against viral infections. Lectin due to non-protein-specific association to insulin receptors, possesses insulin-like activity. Lectin act on peripheral tissues and lower the concentration of blood glucose, and is related to the effect of insulin in brain (Gadang et al. 2011; Xu et al. 2015). Singh et al. (2017) reported that the extract of bitter gourd prevents the regression of blood vessels and granulation tissue, results in improving and accelerating wound healing.

3.5 Concluding Remarks

Bitter gourd has been traditionally used for a number of medicinal purposes since a long time across the globe besides being used as a nutritional food. It is a treasure house of wide array of bioactive compounds, which illustrates its wide application in a number of diseases.

The mechanism of action on various receptors further justifies its usage as a multiple drug therapy. The presence of cucurbitacins besides the presence of alkaloids, proteins, saponins, etc., makes it a unique natural drug.

References

Ahmad N, Hasan N, Ahmad Z, Zishan M, Zohrameena S et al (2016) *Momordica charantia*: for traditional uses and pharmacological actions. J Drug Deliv Therapeut 6(2):40–44

Ahmad Z, Zamhuri KF, Yaacob A, Siong CH, Selvarajah M et al (2012) In vitro anti-diabetic activities and chemical analysis of polypeptide-K and oil isolated from seeds of *Momordica charantia* (bitter gourd). Molecules 17(8):9631–9640

Alam M, Uddin R, Subhan N, Rahman MM, Jain N et al (2015) Beneficial role of bitter melon supplementation in obesity and related complications in metabolic syndrome. J Lipid Res 49:61–69

Alam S, Asad SM, Asdaq M, Prasad SM et al (2009) Antiulcer activity of methanolic extract of *Momordica charantia* L in rats. J Ethnopharmacol 123(3):464–469

Ali L, Khan L, Mamun AK, Mosihuzzaman MI, Nahar N et al (1993) Studies on hypoglycemic effects of fruit pulp, seed, and whole plant of *Momordica charantia* on normal and diabetic model rats. Planta Med 59:408–412

Aljohi A, Matou-Nasri S, Ahmed N (2016) Antiglycation and antioxidant properties of *Momordica Charantia*. PLoS One 11

Aziz M, Karboune S (2016) Natural antimicrobial/ antioxidant agents in meat and poultry products as well as fruits and vegetables. Crit Rev Food Sci Nutr 58(3):486–511

Bai J, Zhu Y, Dong Y (2016) Response of gut microbiota and inflammatory status to bitter melon (*Momordica charantia* L) in high fat diet induced obese rats. J Ethnopharmacol 16:378–387

Bauri R, Tigga M, Kullu S (2015) A review on use of medicinal plants to control parasites. Indian J Nat Prod Resour 6:268–277

Beloin N, Gbeassor M, Akpagana K, Hudson J, de Soussa K et al (2005) Ethnomedicinal uses of *Momordica charantia* (Cucurbitaceae) in Togo and relation to its phytochemistry and biological activity. J Ethnopharmacol 96:49–55

Bhattacharya KS, Muhammad N, Steele R, Peng G, Ray RB (2016) Immunomodulatory role of bitter melon extract in inhibition of head and neck squamous cell carcinoma growth. Oncotarget 22:33202–33209

Biswas AR, Ramaswamy S, Bapna JS (1991) Analgesic effect of *Momordica charantia* seed extract in mice and rats. J Ethnopharmacol 31:115–118

Blaskovich MA, Sun J, Cantor A, Turkson J, Jove R et al (2003) Discovery of JSI-124 (Cucurbitacin I), a selective Janus kinase/ signal transducer and activator of transcription 3 signaling pathway inhibitor with potent antitumor activity against human and murine cancer cells in mice. Can Res 63:1270–1279

Bowman T, Garcia R, Turkson J, Jove R (2000) STATs in oncogenesis. Oncogene 19:2474–2488

Bowman T, Yu H, Sebti S, Dalton W, Jove R (1999) Signal transducers and activators of transcription: novel targets for anticancer therapeutics. Cancer Control 6:427–435

Braca A, Siciliano T, D'Arrigo M, Germanò MP (2008) Chemical composition and antimicrobial activity of *Momordica Charantia* seed essential oil. Fitoterapia 79(2):123–125

Brandao DO, Guimaraes GP, Santos RL (2016) Model analytical development for physical, chemical, and biological characterization of *Momordica Charantia* vegetable. Drug J Anal Meth Chem 75:282–297

Cao H, Sethumadhavan K, Grimm CC, Ullah AH (2014) Characterization of a soluble phosphatidic acid phosphatase in bitter melon (*Momordica Charantia*). PLoS ONE 9(9):106–113

Chao CY, Sung PJ, Wang WH, Kuo YH et al (2014) Anti-inflammatory effect of *Momordica Charantia* in sepsis mice. Molecules 19(8):12777–12788

Chao CY, Yin MC, Huang CJ (2011) Wild bitter gourd extract up-regulates mRNA expression of PPARα, PPARγ and their target genes in C57BL/6 J mice. J Ethnopharmacol 135(1):156–161

Chaturvedi P (2009) Bitter melon protects against lipid peroxidation caused by immobilization stress in albino rats. Intl J Vit Nutr Res 79(1):48–56

Chaturvedi P (2012) Antidiabetic potentials of *Momordica charantia*: Multiple mechanisms behind the effects. J Med Food 15(2):101–107

Chaturvedi P, George S, Milinganyo M, Tripathi YB (2004) Effect of *Momordica charantia* on lipid profile and oral glucose tolerance in diabetic rats. Phytother Res 18:954–956

Chen Q, Li ET (2005) Reduced adiposity in bitter melon (*Momordica charantia*)–fed rats is associated with increased lipid oxidative enzyme activities and uncoupling protein expression. J Nutr 135:2517–2523

Chen J, Tian R, Qiu M, Lu L, Zheng Y, Zhang Z (2008) Trinorcucurbitane and cucurbitane triterpenoids from the roots of *Momordica Charantia*. Photochemistry 69 (4):1043–1048

Chen Q, Chan L, Li E (2003) Bitter melon (*Momordica charantia*) reduces adiposity, lowers serum insulin and normalizes glucose tolerance in rats fed a high fat diet. J Nutr 133:1088–1093

Choi J, Lee KT, Jung H, Park HS (2002) Anti-rheumatoid arthritis effect of the *Kochia scoparia* fruits and activity comparison of momordin lc, its prosapogenin and sapogenin. Arch Pharmacol Res 25:336–342

Chunthorng-Orn J, Panthong S, Itharat A (2012) Antimicrobial, antioxidant activities and total phenolic content of Thai medicinal plants used to treat HIV patients. J Med Ass Thai 95(1):154–158

Dandawate PR, Subramaniam D, Padhye SB, Anant S (2016) Bitter melon: A panacea for inflammation and cancer. Chin J Nat Med 14(2):81–100

Dans AML, Villarruz MW (2007) The effect of *Momordica charantia* capsule preparation on glycemic control in type 2 diabetes mellitus needs further studies. J Clin Epidemiol 60:554–559

Dar UK, Owais F, Ahmad M, Rizwani GH (2014) Biochemical analysis of the crude extract of *Momordica Charantia* (L). J Pharmaceut Sci 27(6):2237–2240

Deep G, Dasgupta T, Rao AR, Kale RK (2004) Cancer preventive potential of *Momordica charantia* L against benzo(a)pyrene induced fore-stomach tumourigenesis in murine model system. J Exp Biol 42:319–322

Dinan L, Harmatha J, Lafont R (2001) Chromatographic procedure for the isolation of plant steroids. J Chromatogr A 935:105–123

Dong Y, Lu B, Zhang X, Zhang J, Lai L et al (2010) Cucurbitacin E, a tetracyclic triterpenes compound from Chinese medicine, inhibits tumor angiogenesis through VEGFR2 mediated JAK2/ STAT3 signaling pathway. Carcinogenesis 31:2097–2104

Duangmano S, Dakeng S (2010) Antiproliferative effects of cucurbitacin B in breast cancer cells: down-regulation of the c-myc/htert/telomerase pathway and obstruction of the cell cycle. Intl J Mol Sci 11:5323–5338

Duncan KL, Duncan MD, Alley MC, Sausville EA (1996) Cucurbitacin E-induced disruption of the actin and vimentin cytoskeleton in prostate carcinoma cells. Biochem Pharmacol 52:1553–1560

Escandell JM, Kaler P, Recio MC, Sasazuki T, Shirasawa S et al (2008) Activated kRas protects colon cancer cells from Cucurbitacin-induced apoptosis: the role of p53 and p21. Biochem Pharmacol 76:198–207

Esterbauer H (1993) Cytotoxicity and genotoxicity of lipid oxidation products. Amer J Clin Nutr 57(Suppl 5):779S–785S

Faden AI, Loane DJ (2015) Chronic neurodegeneration after traumatic brain injury: alzheimer disease, chronic traumatic encephalopathy, or persistent neuroinflammation. Neurotherapeut 12(1):143–150

Fan JM, Zhang Q, Xu J, Zhu S, Ke T et al (2009) Inhibition on hepatitis B virus *in-vitro* of recombinant MAP30 from bitter melon. Mol Biol Rep 36(2):381–388

Fan X, He L, Meng Y, Li G, Li L, Meng Y (2015) A-MMC and MAP30, two ribosome-inactivating proteins extracted from *momordica charantia*, induce cell cycle arrest and apoptosis in a549 human lung carcinoma cells. Mol Med Rep 11(5):3553–3558

Fernandes NPC, Lagishetty CV, Panda VS, Naik SR (2007) An experimental evaluation of the antidiabetic and antilipidemic properties of a standardized *Momordica charantia* fruit extract. BMC Complem Altern Med 7:29–33

Ferrándiz ML, Alcaraz MJ (1991) Anti-inflammatory activity and inhibition of arachidonic acid metabolism by flavonoids agents actions. Springer 32:283–288

Frohne D (1983) A coloured atlas of poisonous plants. Wolf, London

Gadang V, Gilbert W, Hettiararchchy N, Horax R, Katwa L et al (2011) Dietary bitter melon seed increases peroxisome proliferator-activated receptor-γ gene expression in adipose tissue, down-regulates the nuclear factor-κb expression, and alleviates the symptoms associated with metabolic syndrome. J Med Food 14(1–2):86–93

Gbedema SY, Bayor MT, Annan K, Wright CW (2015) Clerodane diterpenes from *Polyalthia longifolia* (Sonn) Thw var pendula: potential antimalarial agents for drug resistant *Plasmodium falciparum* infection. J Ethnopharmacol 169:176–182

Gil MI, Tomás-Barberán FA, Hess-Pierce B, Kader AA (2002) Antioxidant capacities, phenolic compounds, carotenoids, and vitamin c contents of nectarine, peach, and plum cultivars from California. J Agri Food Chem 50:4976–4982

Grover JK, Yadav SP (2004) Pharmacological actions and potential uses of *Momordica charantia*—a review. J Ethnopharmacol 93:123–132

Gupta M, Gautam S, Ajay Y, Bhaduria R (2011) Review a article *Momordica charantia* Linn (Karela): nature' s silent healer. Intl J Pharm Sci Rev Res 11(1):32–37

Gupta S, Raychaudhuri B, Banerjee S, Mukhopadhaya S, Datta SC (2010) Momordicatin purified from fruits of *Momordica charantia* is effective to act as a potent antileishmania agent. Parasitol Intl 59(2):192–197

Harinantenaina L, Tanaka M, Takaoka S, Oda M, Mogami O et al (2006) *Momordica charantia* constituents and antidiabetic screening of the isolated major compounds. Chem Pharm Bull 54:1017–1021

Higashio H (2002) Value adding technologies to commodies in vegetable production. Res J Food Agri 25:8–22

Hrckova G, Velebny S (2013) Pharmacological potential of selected natural compounds in the control of parasitic diseases brief in pharmaceutical science and drug development: parasitic Helminths of humans and animals: Health impact and control. Springer, Vienna, pp 29–99

Huang Z, Huang Q, Ji L, Wang Y, Qi X, Liu L et al (2016) Epigenetic regulation of active Chinese herbal components for cancer prevention and treatment: a follow-up review. Pharmacol Res 114:1–12

Hussan F, Teoh SL, Muhamad N, Mazlan M, Latiff AA (2014) Momordica charantia ointment accelerates diabetic wound healing and enhances transforming growth factor-β expression. J Wound Care 23(8):404–407

Iglesias MA, Ye JM (2002) AICAR administration causes an apparent enhancement of muscle and liver insulin action in insulin-resistant high-fat-fed rats. Diabetes 51:2886–2894

Ilhan M, Bolat IE (2015) Topical application of olive oil macerate of momordica charantia l promotes healing of excisional and incisional wounds in rat buccal mucosa. Arch Oral Biol 60(12):1708–1713

Iqbal Z, Mufti K, Khan M (2004) Anthelmintic effects of condensed tannins. Intl J Agri Biol 4:438–440

Jain P, Singh S, Singh S, Verma S, Kharya M, Solanki S (2013) Anthelmintic potential of herbal drugs international. J Res Dev 2:412–417

Jayaprakasam B, Seeram NP, Nair MG (2003) Anticancer and anti-inflammatory activities of cucurbitacins from Cucurbita andreana. Cancer Lett 189:11–16

Jianchao C, Renrong T, Qiu M, Lu L, Yongtang Z, Zhang Z (2008) Trinorcucurbitane and cucurbitane triterpenoids from the roots of Momordica charantia. Phytochemistry 69:1043–1048

Jorn G, Inge S, Hans CA (2006) Cucurbitacins in plant food. TemaNord 2006:556

Kai H, Akamatsu E, Torii E, Kodama E, Yukizaki H, Matsuno K et al (2011) Inhibition of proliferation by agricultural plant extracts in seven human adult T-cell leukaemia (atl)-related cell lines. J Nat Sci 65(3–4):651–655

Kamal R, Yadav S, Mathur M, Katariya P (2011) Antiradical efficiency of 20 selected medicinal plants. J Nat Prod 26(11):1054–1062

Kamath CC, Vickers KS, Ehrlich A, McGovern L, Johnson J et al (2008) Behavioral interventions to prevent childhood obesity: a systematic review and meta analyses of randomized trials. J Clin Endo Meta 93(12):4606–4615

Kappagoda S, Singh U, Blackburn BG (2011) Antiparasitic therapy. Mayo Clin Proc 86:561–583

Kavitha N, Babu SM, Rao ME (2011) Influence of Momordica charantia on oxidative stress-induced perturbations in brain monoamines and plasma corticosterone in albino rats. Indian J Pharmacol 43(4):424–428

Kee HC, Hongtao X (2008) Methods of inducing apoptosis in cancer treatment by using cucurbitacins. US2008/0207578A1: 28

Kobori M, Nakayama H, Fukushima K, Ohnishi-Kameyama M et al (2008) Bitter gourd suppresses lipopolysaccharide-induced inflammatory responses. J Agri Food Chem 56:4004–4011

Konishi T, Satsu H, Hatsugai Y, Aizawa K, Inakuma T, Nagata S et al (2004) Inhibitory effect of a bitter melon extract on the p-glycoprotein activity in intestinal caco-2 cells. Br J Pharmacol 143:379–387

Kumar KPS, Bhowmik D (2010) Traditional medicinal uses and therapeutic benefits of Momordica charantia Linn. Intl J Pharmaceut Sci Rev Res 4(3):23–28

Laverdure S, Polakowski N, Hoang K, Lemasson I (2016) Permissive sense and antisense transcription from the 5′ and 3′ long terminal repeats of human t-cell leukemia virus type 1. J Virol 90(7):3600–3610

Li Y, Xu C, Yang XJ, Wu XG et al (2016) One New 19-nor cucurbitane-type triterpenoid from the stems of Momordica charantia. Nat Prod Res 30(8):973–978

Limtrakul P, Khantamat O, Pinth K (2004) Inhibition of p-glycoprotein activity and reversal of cancer multidrug resistance by Momordica charantia extract. Cancer Chemother Pharmacol 54:525–530

Lin J, Tang C (2008) Strawberry, loquat, mulberry, and bitter melon juices exhibit prophylactic effects on LPS-induced inflammation using murine peritoneal macrophages. Food Chem 107:1587–1596

Liu T, Zhang M, Zhang H, Sun C, Deng Y (2000) Inhibitory effects of Cucurbitacin B on laryngeal squamous cell carcinoma. Eur Arch Otorhinolaryngol 265:1225–1232

Liu C, Cai D, Zhang L, Tang W, Yan R, Guo H et al (2016) Identification of hydrolyzable tannins (punicalagin, punicalin and geraniin) as novel inhibitors of hepsatitis B virus covalently closed circular DNA. Antiviral Res 134:97–107

Liu JQ, Chen JC, Wang CF, Qiu MH (2009) New Cucurbitane triterpenoids and steroidal glycoside from Momordica charantia. Molecules 14(12):4804–4813

Liu XR, Deng ZY, Fan YW, Li J, Liu ZH (2010) Mineral elements analysis of Momordica charantiap seeds by ICP-AES and fatty acid profile identification of seed oil by GC-MS. Guang Pu Xue Yu Guang Pu Fen Xi 30(8):2265–2268

Martínez-Abundis E, Mendez-Del Villar M, Perez R et al (2016) Novel nutraceutic therapies for the treatment of metabolic syndrome. World J Diabetes 7(7):142–152

Maurya P, Sharma P, Mohan L, Batabyal L, Srivastava CN (2009) Evaluation of larvicidal nature of fleshy fruit wall of Momordica charantia Linn (family: cucurbitaceae) in the management of mosquitoes. J Parasitol Res 105(6):1653–1659

Melzig MF, Bader G, Loose R (2001) Investigations of the mechanism of membrane activity of selected triterpenoid saponins. Planta Med 67:43–48

Meng Y, Lin S, Liu S, Fan X, Li G, Meng YA (2014) Novel method for simultaneous production of two ribosome-inactivating proteins, α-mmc and map30, from Momordica charantia L. PLoS ONE 9(7):101–109

Mesia GK, Tona GL, Nanga TH, Cimanga RK, Apers S, Cos P et al (2008) Antiprotozoal and cytotoxic screening of 45 plant extracts from democratic republic of congo. J Ethnopharmacol 115(3):409–415

Mwambete KD (2009) The In-vitro antimicrobial activity of fruit and leaf crude extracts of momordica *charantia*: a tanzania medicinal plant. Afr Health Sci 9 (1):34–39

Nagasawa H, Watanabe K, Inatomi H (2002) Effects of bitter melon (*Momordica charantia* L) or ginger rhizome (*Zingiber offifinale*) on spontaneous mammary tumorigenesis in SHN mice. Amer J Chin Med 30:195–205

Namsa ND, Mandal M, Tangjang S, Mandal SC (2011) Ethnobotany of the monpa ethnic group at Arunachal Pradesh, India. J Ethnobiol Ethnomed 7:31–37

Nerurkar PV, Lee YK, Nerurkar VR (2010) *Momordica charantia* (Bitter Melon) inhibits primary human adipocyte differentiation by modulating adipogenic genes. BMC Complem Altern Med 10:34

Nhiem NX, Yen PH, Ngan NT, Quang TH, Van-Kiem P et al (2012) Inhibition of nuclear transcription factor-κB and activation of peroxisome proliferator-activated receptors in Hepg2 cells by cucurbitane-type triterpene glycosides from *Momordica charantia*. J Med Food

Olasehinde GI, Ojurongbe O, Adeyeba A, Fagade O et al (2014) In-vitro studies on the sensitivity pattern of *Plasmodium falciparum* to anti-malarial drugs and local herbal extracts. Malaria J 13:63

Padmashree A, Sharma GK, Semwal AD, Bawa AS (2011) Studies on the antioxygenic activity of bitter gourd (*momordica charantia*) and its fractions using various *In-vitro* models. J Sci Food Agri 91(4):776–782

Pandit S, Kanjilal S, Awasthi A, Chaudhary A, Banerjee D, Bhatt BN et al (2016) Evaluation of herb-drug interaction of a polyherbal ayurvedic formulation through high throughput cytochrome p450 enzyme inhibition assay. J Ethnobiol Ethnopharmacol 16:304–308

Park CS, Lim H, Han KJ, Baek SH, Sohn HO, Lee DW et al (2004) Inhibition of nitric oxide generation by 23,24-dihydrocucurbitacin D in mouse peritoneal macrophages. J Pharmacol Exp Therapeut 309:705–710

Phillips EA, Sexton DW, Steverding D (2013) Bitter melon extract inhibits proliferation of *Trypanosoma brucei* bloodstream forms in vitro experimental. Parasitology 133(3):353–356

Pongnikorn S, Fongmoon D, Kasinrerk W Limtrakul (2003) Effect of bitter melon (*Momordica charantia* Linn) on level and function of natural killer cells in cervical cancer patients with radiotherapy. J Med Assoc Thai 86:61–68

Prabhakar K, Kumar LS, Rajendran S, Chandrasekaran M, Bhaskar K, Sajit-Khan AK (2008) Antifungal activity of plant extracts against candida species from oral lesions. Indian J Pharma Sci 70(6):801–803

Puri R, Sud R, Khaliq A, Kumar M, Jain S (2011) Gastrointestinal toxicity due to bitter bottle gourd (*Lagenaria Siceraria*)—A report of 15 cases. J Gastroenterol 30(5):233–236

Qader SW, Abdulla MA, Chua LS, Najim N, Zain MM, Hamdan S (2011) Antioxidant, total phenolic content and cytotoxicity evaluation of selected Malaysian plants. Molecules 16(4):3433–3443

Rajeswari V (2014) Anthelmintic activity of plants: a review. Res J Phytochem 8:57–63

Raman A, Lau C (1996) Anti-diabetic properties and phytochemistry of *Momordica charantia* L (Cucurbitaceae). Phytomedicine 2(4):349–362

Rawlings CR, Fremlin GA, Nash J, Harding KA (2016) Rheumatology perspective on *Cutaneous vasculitis*: assessment and investigation for the non-rheumatologist. Intl Wound J 13(1):17–21

Rehm S (1957) Bitter principles of the Cucurbitaceae VII the distribution of bitter principles in this plant family. J Sci Food Agri 8:679–686

Saba AB, Oridupa AO (2010) Search for a novel antioxidant, anti-inflammatory/analgesic or anti-proliferative drug: Cucurbitacins. J Med Plants Res 4:2821–2826

Sabourian R, Karimpour-Razkenari E, Saeedi M, Bagheri MS et al (2016) Medicinal plants used in Iranian traditional medicine as contraceptive agents. Curr Trends Biotechnol Phar 17(11):974–985

Sagor AT, Chowdhury MR, Tabassum N, Hossain H, Rahman MM, Alam MA (2015) Supplementation of fresh ucche (*momordica charantia* l var *muricata* willd) prevented oxidative stress, fibrosis and hepatic damage in ccl4 treated rats. BMC Complem Altern Med 15:115

Santos KK, Matias EF, Sobral-Souza CE, Tintino SR, Morais B et al (2012) Trypanocide, cytotoxic, and antifungal activities of *Momordica charantia*. Pharm Biol 50(2):162–166

Sathishsekar D, Subramanian S (2005) Beneficial effects of *Momordica charantia* seeds in the treatment of STZ-induced diabetes in experimental rats. Biol Pharm Bull 28:978–983

Sato M, Ueda T, Nagata K, Shiratake S, Tomoyori H, Kawakami M et al (2011) Dietary Kakrol (*Momordica Dioica* Roxb.) flesh inhibits triacylglycerol absorption and lowers the risk for development of fatty liver in rats. Exp Biol Med 236(10):1139–1146

Senanayake GV, Maruyama M, Sakono M, Fukuda N, Morishita T, Yukizaki C, Kawano M, Ohta H (2004) The effects of bitter melon (*Momordica charantia*) extracts on serum and liver lipid parameters in hamsters fed cholesterol-free and cholesterol-enriched diets. J Nutr Sci Vitaminol 50:253–257

Shodehinde SA, Adefegha SA, Oboh G, Oyeleye SI, Olasehinde TA et al (2016) Phenolic composition and evaluation of methanol and aqueous extracts of bitter gourd (*momordica charantia* l) leaves on angiotensin-i-converting enzyme and some pro-oxidant-induced lipid peroxidation *in vitro* evidence–based. Complem Altern Med 21(4):67–76

Singh A, Singh SP, Bamezai R (1998) *Momordica charantia* (bitter gourd) peel, pulp, seed and whole fruit extract inhibits mouse skin papillomagenesis. Toxicol Lett 94:37–46

Singh R, Kumar A, Singh ML, Maurya SK, Pandey KD (2017) Microbial diversity in the rhizosphere of *Momordica charantia* L (bitter gourd). Intl J Curr Micro Appl Sci 6(2):67–76

Sun J, Blaskovich MA, Jove R, Livingston SK, Coppola D, Sebti SM, Cucurbitacin Q (2005) A selective STAT3 activation inhibitor with potent antitumor activity. Oncogene 24:3236–3245

Tabackman AA, Frankson R, Marsan ES, Perry K, Cole KE (2016) Structure of 'linkerless' hydroxamic acid inhibitor-HDAC8 complex confirms the formation of an isoform-specific subpocket. J Struc 195 (3):373–378

Tambor M, Pavlova M, Golinowska S, Arsenijevic J, Groot W (2016) Financial incentives for a healthy life style and disease prevention among older people a systematic literature review. Health Serv Res 16 (5):426

Tan MJ, Ye JM, Turner N, Hohen Behrens C, Ke CQ, Tang CP et al (2008) Antidiabetic activities of triterpenoids isolated from bitter melon associated with activation of the AMPK pathway. Chem Biol 15:263–273

Tannin-Spitz T, Bergman M, Grossman S (2007) Cucurbitacin glucosides: antioxidant and free-radical scavenging activities. Biochem Biophys Res Commun 364:181–186

Thenmozhi AJ, Subramanian P (2012) Antioxidant potential of *Momordica charantia* in ammonium chloride-induced hyperammonemic rats. J Evid-Based Complem Altern Med 2011:1–7

Turkson J, Jove R (2000) STAT proteins: novel molecular targets for cancer drug discovery. Oncogene 19 (66):13–26

Veerakumari L (2015) Botanical anthelmintics. Asian J Sci Technol 6:1881–1894

Virdi J, Sivakami S, Shahani S, Suthar AC, Banavalikar MM et al (2003) Antihyperglycemic effects of three extracts from *Momordica charantia*. J Ethnopharmacol 88:107–111

Wang BL, Zhang WJ, Zhao J, Wang FJ, Fan LQ, Wu YX, Hu ZB (2010) Gene cloning and expression of a novel hypoglycemic peptide from *Momordica charantia*. J Sci Food Agri 91(13):2443–2448

Welihinda J, Karunanayake EH (1986) Extra-pancreatic effects of *Momordica charantia* in rats. J Ethnopharmacol 17:247–255

Williams A, Fryganas C, Ramsay A, Mueller-Harvey I, Thamsborg S (2014) Direct anthelmintic effects of condensed tannins from diverse plant sources against *Ascaris suum*. PLoS ONE 9:997

Wink M (2012) Medicinal plants: a source of anti-parasitic secondary metabolites. Molecules 17:12771–12791

Xu X, Shan B, Liao CH, Xie JH, Wen PW, Shi JY (2015) Anti-diabetic properties of *Momordica charantia* l polysaccharide in alloxan-induced diabetic mice. Intl J Biol Sci 81:538–543

Yadav BS, Yadav R, Yadav RB, Garg M (2016) Antioxidant activity of various extracts of selected gourd vegetables. J Food Sci 53(4):1823–1833

Yaldız G, Sekeroglu N, Kulak M, Demirkol G (2015) Antimicrobial activity and agricultural properties of bitter melon (*Momordica charantia* l) grown in northern parts of Turkey: a case study for adaptation. Nat Prod Res 29(6):543–545

Yang SJ, Choi JM, Park SE, Rhee EJ, Lee WY et al (2015) Preventive effects of bitter melon (*Momordica charantia*) against insulin resistance and diabetes are associated with the inhibition of NF-KB and JNK pathways in high-fat-fed oletf rats. J Nutr Biochem 26 (3):234–240

Yasui Y, Hosokawa M, Sahara T, Suzuki R, Ohgiya S, Kohno H, Tanaka T, Miyashita K (2005) Bitter gourd seed fatty acid rich in 9c,11t,13t-conjugated linolenic acid induces apoptosis 196 S P TAN ET AL and up-regulates the GADD45, p53 and PPARγ in human colon cancer Caco-2 cells. Prostaglandins Leukotrienes Essent Fatty Acids 73:113–119

Ye JM, Ruderman NB, Kraegen EW (2005) AMP-activated protein kinase and malonyl-CoA: targets for treating insulin resistance. Drug Discov Today Therap Strateg 2:157–163

Yehye WA, Abdul RN, Saad O, Ariffin A, Hamid SB et al (2016) Rational design and synthesis of new, high efficiency, multipotent schiff base-1,2,4-triazoleantioxidants bearing butylated hydroxytoluene moieties. Molecules 21(7):84–87

Yoon YA, Kim H, Lim Y, Shim YH (2006) Relationships between the Larval growth inhibition of *Caenorhabditis elegans* by apigenin derivatives and their structures archives. Pharma Res 29:582–586

Yuan G, Mark LW, Guoqing H, Min Y, Li D (2006) Natural products and anti-inflammatory activity. Asia Pac J Clin Nutr 15:143–152

Zhu ZJ, Zhong ZC, Luo ZY, Xiao ZY (1990) Studies on the active constituents of *Momordica charantia* L. Yao Xue Xue Bao 25:898–903

Zhu F, Zhang P (2013) Alpha-Momorcharin, a RIP produced by bitter melon, enhances defense response in tobacco plants against diverse plant viruses and shows antifungal activity *in-vitro*. Planta 237(1): 77–88

Genetic Resources and Genetic Diversity in Bitter Gourd

Tusar Kanti Behera, Shyam Sundar Dey, Sutapa Datta and Chittaranjan Kole

Abstract

The genus *Momordica* comprises important group of vegetables including bitter gourd which is a widely cultivated species in different parts of the world. Most of the species under this genus are distributed in Asia, and a few are domesticated in the African region. Among the 45 species of *Momordica* reported worldwide, six are found in India. Understanding the species diversity, gene pool and genetic resources are keys to utilize the desirable traits present in the different wild relatives and feral forms. The primary gene pool of *M. charantia* consists of different variants under this species, besides *M. balsamina*. It is possible to derive a few fertile hybrids through crossing of *M. charantia* with *M. balsamina* because of their same chromosome number. The secondary gene pool consists of other species of *Momordica*, which may produce a few fertile hybrids through pollination involving a large number of flower buds. The tertiary gene pool consists of other species of *Momordica* and different other genera of Cucurbitaceae from where it is possible to transfer the desirable traits using special tools like protoplast fusion. In the species *M. charantia*, a huge diversity is reported for important morphological traits including fruit shape, size, yield, and earliness besides several other vegetative and commercial traits. The genus *Momordica* is known for its antidiabetic and other medicinal properties. Huge diversity is reported in terms of concentration of different phytomedicinal compounds including momordicin, momorcharin, β-carotene, and lycopene. While studying diversity in molecular level, RAPD, ISSR, SSR, and other sequence-based markers were used that revealed the molecular diversity. The combined information available in terms of morphological, nutritional, and molecular diversity present in this important crop can help the breeder in deciding the breeding material and approach to harness more productivity and enhance nutritional traits in bitter gourd using different methods.

4.1 Introduction

This genus, *Momordica*, includes 45 species domesticated in Asia and Africa (Robinson and Decker-Walters 1999). There is enormous diversity present in the genus, and several desirable and novel traits are present in different feral forms and wild relatives. Proper understanding regarding the

T. K. Behera (✉) · S. S. Dey
Division of Vegetable Sciences, ICAR-Indian Agricultural Research Institute, Pusa, New Delhi 110012, India
e-mail: tusar@rediffmail.com

S. Datta · C. Kole
ICAR-National Institute for Plant Biotechnology, Pusa, New Delhi 110012, India

© Springer Nature Switzerland AG 2020
C. Kole et al. (eds.), *The Bitter Gourd Genome*, Compendium of Plant Genomes,
https://doi.org/10.1007/978-3-030-15062-4_4

species diversity and interrelation among the different species is vital to exploit the genetic diversity present in different other wild relatives and their possible introgression into the cultivated genotypes of *Momordica*. Therefore, gene pool will provide an idea about the kind of diversity and the strategies to be adopted for their successful utilization in the improvement programs of bitter gourd.

4.2 Gene Pools in *Momordica*

Bitter gourd (*Momordica charantia* L.) is the most widely cultivated species of the genus *Momordica* and is grown in India, Sri Lanka, Philippines, Thailand, Malaysia, China, Japan, Australia, tropical Africa, South America, and the Caribbean. Bitter gourd is consumed regularly as part of several Asian cuisines and has been used for centuries in ancient traditional Indian, Chinese, and African pharmacopoeia. Other *Momordica* species, apart from their importance as wild relatives of bitter gourd, have direct utility as nutritious vegetables and multipurpose medicinal plants and can be exploited as alternative crops. De Wilde and Duyfjes (2002) have reported ten species from South East Asia, of which six each occur in Malaysia and India, where *M. balsamina* L., *M. charantia* L., *M. subangulata* Blume subsp. *renigera* (G. Don) W. J. de Wilde, and *M. cochinchinensis* (Lour.) Spreng. are most common. A few more species have been described later by different workers (De Wilde and Duyfjes 2002; Jongkind 2002; Joseph and Antony 2007). *Momordica* species, namely *M. charantia*, *M. balsamina*, *M. dioica*, *M. cymbalaria*, *M. denudata*, *M. macrophylla*, *M. subangulata*, and *M. cochinchinensis*, were reported to occur in India (Miniraj et al. 1993). Among these, *M. macrophylla* is treated as synonymous with *M. cochinchinensis* (Jeffrey 1980, 2001; De Wilde and Duyfjes 2002). *M. cymbalaria* Fenzl ex Naud. (syn. *M. tuberose* (Roxb.)) was transferred to the genus *Luffa* (Chakravarty 1959), and the occurrence/ existence of *M. denudata* in India is doubtful (Joseph 2005). According to the latest revision of

Indian *Momordica* spp., there are six species found in different parts of India out of which four are dioecious and two are monoecious (Joseph 2005). An exchange of genes between the cultivated and feral or semi-feral species of *Momordica* would have immense potential to generate more genetic divergence and introgression of novel and desirable traits present in different wild species. The identification and incorporation of resistance to economically important pests such as fruit fly and various foliar pathogens are important for bitter gourd improvement.

Gene pool of any crop is collection of all the genes available for breeding use which include all cultivars, wild species, and wild relatives of a particular species. The word gene pool was formulated first in the 1920s as *genofond* (gene fund), a word that was adopted in USA from the Soviet Union by Theodosius Dobzhansky, who translated it into English as 'gene pool.' Based on the degree of relationship, the gene pool of any crop can be divided into three groups, (i) primary gene pool, (ii) secondary gene pool, and (iii) tertiary gene pool (Harlan and Wet 1971).

4.2.1 Primary Gene Pool of *Momordica*

The gene pool which leads to the production of fertile hybrids is known as primary gene pool (GP1). It includes plants of the same species or of closely related species. In such gene pool, genes can be exchanged between lines simply by making conventional crossing and backcross-assisted breeding. This is also known as gene pool and divided further into subspecies A: cultivated races and subspecies B: spontaneous races (wild or weedy). This is the material of prime breeding importance. In *Momordica*, Group A consists of *M. charantia* (var. *charantia* and var. *muricata*) and *M. balsamina*. The cultivated *M. charantia* var. *charantia* crosses with the semidomesticated/ wild bitter gourd (*M. charantia* var. *muricata*), and gene exchange occurs freely within the complex. Interspecific crosses were not successful between *M. charantia* and *M. balsamina*

(Joseph 2005). However, Singh (1990) obtained two fruits from the cross *M. charantia* × *M. balsamina* after 200 pollinations. *M. charantia* and *M. balsamina* had almost the same number of median and submedian chromosomes and had fairly strong karyomorphological similarities (Trivedi and Roy 1972; Singh 1990). A high bivalent frequency with a normal meiotic cycle in the progeny of *M. balsamina* × *M. charantia* indicated that the two taxa are phylogenetically close (Singh 1990). The monoecious species *M charantia* and *M. balsamina* produce edible fruits and have been widely distributed as crops becoming naturalized throughout the tropics.

4.2.2 Secondary Gene Pool of *Momordica*

The genetic material that leads to partial fertility on crossing with GP1 is referred to as secondary gene pool. Species of this group are closely related and can cross and produce at least some fertile hybrids. As would be expected by members of different species, there are some reproductive barriers between members of the primary and secondary gene pools: GP2 consists of *M. dioica*, *M. sahyadrica*, *M. subangulata* subsp. *renigera*, and *M. cochinchinensis*. Bharathi (2010) reported successful production of viable seeds between all combinations of dioecious group except two combinations *M. cochinchinensis* × *M. dioica* and *M. cochinchinensis* × *M. sahyadrica*. Complete incompatibility was also observed between monoecious and dioecious species. A very high seed set and germination percentage were observed between the crosses of *M. dioica* and *M. sahyadrica*. It was interesting to note that when *M. dioica* was used as female parent the success rate was 75% while success rate of 80.95% was recorded when *M. sahyadrica* was used as female parent. Significantly lower seed set as well as germination rate was evident when a diploid was used as female parent, while a good seed set and germination percentage were observed whenever tetraploid species (*M. subangulata* subsp. *renigera*) was used as female parent. It was observed that *M. subangulata* subsp. *renigera* (tetraploid) recorded higher fruit set

when crossed with diploid species *M. sahyadrica* and *M. dioica* (84% and 65%, respectively) while in reciprocal cross of *M. sahyadrica* and *M. dioica* with *M. subangulata* subsp. *renigera* the fruit set was 50% and 53.33%, respectively, and they also showed very low germination percentage (20% and 10%, respectively). Further, *M. cochinchinensis* (diploid) as a female parent was crossable only with *M. subangulata* subsp. *renigera* (tetraploid) but a very low fruit set (2%) as well as germination percentage (20%) was observed.

4.2.3 Tertiary Gene Pool of *Momordica*

Members of this gene pool are more distantly related to the members of the primary gene pool. The primary and tertiary gene pools cannot be intermated, and gene transfer between them is impossible without the use of special nonconventional tools and techniques. This GP3 contains *M. cymbalaria* which did not set any fruits when crossed with rest of the species of the genus *Momordica*. Transfer of gene from such material to primary gene pool is possible with the help of special techniques such as embryo rescue (or embryo culture), induced polyploidy (chromosome doubling), somatic hybridization, genetic engineering, and bridging crosses (e.g., with members of the secondary gene pool).

4.3 Genetic Diversity Study in Bitter Gourd

Genetic diversity is the key in determining potentiality of germplasm and its efficient utilization in crop improvement program. Population with high level of genetic divergence is an important resource for expansion of the genetic base in any crop breeding program. We will discuss the extent of genetic diversity available in bitter gourd at morphological, nutritional, and molecular level. We will also discuss how the available divergence can be used effectively in improvement of bitter gourd. Before going into the available diversity in this crop, it is important to understand the evolution of this crop and important regions of the world where this crop is originated and domesticated over the past.

Momordica charantia L. (Cucurbitaceae), known in English as bitter melon, balsam pear, bitter cucumber, or African cucumber, belongs to a very diverse genus with more than 45 species, and most of them are indigenous to the Old World. Both the cultivated types and wild forms are found naturally in the flora of tropical Africa and Asia. In the New World tropics, it first arrived in Brazil through the slave trade from Africa and later on was distributed into 'Middle America' (Ames 1939). In the Pacific regions, this species is listed as an introduced species in the Cook Islands (Whistler 1990). However, based on early collections by Cook, the cultivated species *M. charantia* is considered to be indigenous to Fiji and the South Pacific but it was also considered as an aboriginal introduction (Smith 1981). The plant of *M. charantia* is herbaceous in nature and sometimes behaves as perennial vine with yellow axillary flowers that are subtended by a bract on the pedicel with abundant pollen production in the male flowers. The leaves are typically pelmet type with 5–9 lobes depending on the genotypes. The fruits are berry type and amenable to dispersal through different animals.

The fruits are green in color with varying degree of bitterness in the pericarp. After attaining maturity, the pericarp becomes yellow and splits open to display bright red, sweet, fleshy arils surrounding the seed. This attractive dispersal mechanism undoubtedly explains the spread of the wild-type wherever it has been accidentally or intentionally introduced by man.

4.3.1 Morphological Diversity

The most important prerequisite for improvement in any crop is to study its germplasm and estimate the extent of variability available. The mode of inheritance of quantitative characters is highly complex and can be understood through the study of genetic parameters such as variability, heritability, and genetic advance in conjunction. Robinson (1966) suggested the need to partition the total variability into genotypic and environmental variance, which helps in selection of desirable genotypes for their use in trait-specific breeding strategies. *Momordica charantia* shows immense diversity for different morphological traits. Sex form is one of the most important traits in bitter gourd which determine the different productivity and yield criteria of a particular genotype. The primary sex form is monoecious with male and female flowers in different nodes in majority of the genotypes. However, gynoecious sex form has been reported from India, Japan, and China (Ram et al. 2002; Behera et al. 2006; Iwamoto and Ishida 2006). In bitter gourd, gynoecism is under the control of a single recessive gene (*gy-1*) (Ram et al. 2006; Behera et al. 2009; Matsumura et al. 2014), whereas two pairs of genes were reported by Cui et al. (2018). Besides sex form, other flowering traits like days to first pistillate flower appearance, node at first pistillate flower appearance, and staminate: pistillate (♂:♀) flower ratio (sex ratio) are directly related to earliness and fruit yield.

Fruits of wild plants are smaller in size with a length of 2–7 cm, generally are pointed at both ends, and have small gray to nearly black seeds. Fruits of the domesticated and cultivated species have varying shape, size, and color. There are genotypes with fruit length up to 60 cm and flattened at the proximal end with larger brown seeds (Walters and Decker-Walters 1988; Yang and Walters 1992). Availability of moderate diversity for different morphological traits in this crop provides ample opportunity to improve various traits. However, understanding the basis of the existing genetic variance and genetics of the economically important traits is prerequisite for their utilization in improvement program.

Studies on morphological divergence in bitter gourd started in the mid-twentieth century, and one of the earliest workers, Srivastava et al. (1976), reported significant diversity for important traits like number of fruits per plant, yield per plant, average weight of the fruits, length of the fruit, number of female flowers per plant, girth of fruit, and days taken for appearance of the first female flower. Earliness is an important trait in bitter gourd, and lowest the node number where first female flower appears the earliest is the genotype (Dey et al. 2006). Therefore, it is important to group the genotypes based on the node number for first female flower appearance to present a better picture regarding maturity duration in bitter gourd. Significant variation for earliness-related traits in bitter gourd has been documented by several workers (Dey et al. 2009; Khan et al. 2015). Besides earliness, yield is another most important trait in bitter gourd. Yield is extremely complex in nature; however, several associated traits like number of female flowers per plant, number of fruits per plant, size and weight of fruits, plant vigor, and vine length are important parameters to determine total yield of a particular genotype (Iqbal et al. 2016; Singh et al 2017; Tiwari et al. 2018). Sahni et al. (1987) in a work with 22 bitter gourd lines reported considerable variability for almost all the characters, which was highest for yield per vine and lowest for fruit diameter. The genotypic coefficient of variation (GCV) was highest for number of fruits per vine, followed by number of female flowers per vine, average fruit weight, and node number for the appearance of the first female flower. The phenotypic coefficient of variation (PCV) also followed the same trend. Bhave et al. (2003) reported higher PCV than the GCV for vine length, branch number per vine, flowering duration, harvesting span, fruit length, average fruit weight, seed number per fruit, fruit number per vine, dry matter per vine, biological yield per vine, and total fruit yield per vine, and the GCV and PCV were higher for branch number per vine, total fruit yield per vine, and vine length in bitter gourd genotypes. Ram et al. (2006) reported maximum coefficients of variation for days to male flower emergence, yield per plant, fruit weight, and fruit length among 12 different traits studied in bitter gourd. Agasimani et al. (2008) observed that the genotypic variation for fruit yield, numbers of fruit per vine, and fruit length was significant for all the three traits in the 12 genotypes of bitter gourd under study. Yadav et al. (2008) conducted variability study in 28 accessions of bitter gourd and reported the maximum fruit length, fruit width, yield per vine, yield per plot, and yield per ha in line MC-84.

The highest number of fruits per vine was recorded in GY-I, and minimum powdery mildew infestation was observed in JMC-22. Dey et al. (2009) reported high PCV and GCV for average fruit weight, average fruit diameter, average flesh thickness, and fruit yield per plant in bitter gourd. PCV was higher than GCV for all the traits studied indicating influence of environment in the expression of traits. Kole et al. (2010) studied 22 genotypes from two different groups of bitter gourd (*M. charantia* var. *charantia* and *M. charantia* var. *muricata*) for important agronomic traits. This collection exhibited highly contrasting plant growth features and fruit attributes including fruit size, color, luster, shape, and surface. The results indicated precise selection of genotypes that represented diverse geographical origins and indicated potential utility of association mapping even if the population size was small. The *muricata* genotypes had small-sized fruits which are green, dull, round, and muricated (spiny), whereas the charantia genotypes did not have any common phenotypic combination and they varied widely among themselves. The results also indicated the possibility of selection of genotypes with contrasting fruit characters as parents for traditional hybridization efforts and also for developing segregating populations as required for molecular mapping of genes/quantitative trait loci (QTLs) controlling fruit traits. Dalamu et al. (2012) studied genetic diversity of 50 indigenous and exotic accessions of bitter gourd. Based on field evaluation over two years, a wide range of variation among the genotypes was observed for yield-related traits which will be useful for selecting the best genotypes with high yield. Khan et al. (2015) analyzed variability among 17 genotypes of bitter gourd for yield and yield-contributing characters. They found a great deal of variation for all the characters among the genotypes. Considering genetic parameters, high GCV was observed for branches per vine, yield per plant, and number of fruit per plant whereas low GCV was observed for days to the first male and female flowering. In all the cases, it was found that PCV was greater than GCV. The highest GCV and PCV values were observed for branch per vine, fruit length, fruit weight, and number of fruit plant which indicated a wide variability among the genotypes and offered better scope of selection. Singh et al. (2017) evaluated genetic diversity of 20 strains of bitter gourd under subtropical conditions of Garhwal Himalayas. The analysis of variance revealed significant differences among the genotypes for almost all the traits (days taken to opening first male and female flower, node number bearing the first male and female flower, days to first fruit harvest, yield per vine, fruit diameter, and fruit weight) except number of nodes per vine and fruit length, and showed adequate genetic variability in the different genotypes. Tiwari et al. (2018) evaluated nine parental lines of bitter gourd for yield and yield-attributing traits. High genotypic as well as phenotypic coefficients of variability were observed for fruit yield per plant and number of fruits per plant. The phenotypic coefficients of variability were only higher for fruit yield per plant. Moderate variability showed in fruit yield per plant and rest of the characters exhibited low coefficient of variability.

4.3.2 Nutritional Diversity with Regard to Medicinal Properties in Bitter Gourd

Bitter gourd is a crop in which all plant parts are used for multiple purposes in different regions of the world and consumed in different ways. The leaves of the plants along with young stems, immature fruits and flowers are consumed as a vegetable in most parts of India and several other regions of Asia (Reyes et al. 1994). This crop is considered to be superior in terms of its nutritional values and medicinal properties among different cucurbitaceous vegetable crops (Reyes et al. 1994). This species is considered as one of the most important vegetable crops with immense medicinal properties, and all parts of the plant are known for phytomedicinal properties in different regions of Asia (Perry 1980), West Africa (Burkhill 1935), and India (Decker-Walters 1999). In the recent past, this species has gained immense importance for its

nutritional traits and phytomedicinal properties in other parts of the world too. This crop is known for its antidiabetic properties throughout the world (Khajuria and Thomas 1993; Platel and Srinivasan 1995; Raman and Lau 1996). Coe and Anderson (1996) documented use of this species in Mexico, Belize, Honduras, Costa Rica, Panama, and the Caribbean, as well as Fiji and Sri Lanka. One feral form of this species is reported for its aphrodisiac property in Mexico (Lira and Caballero 2002). The different parts of the plant, in particular fruits and seeds, contain about 60 bioactive compounds (Raman and Lau 1996) with proven medicinal properties and actions (Ng et al. 1992; Raman and Lau 1996; Basch et al. 2003; http://www.rain-tree.com/bitmelon.htm).

Humans have selected for non-bitter mutants in different species of the Cucurbitaceae in the process of domestication. However, in bitter gourd all the domesticated and the wild-type have bitter fruits. Therefore, it is not certain whether this is because of the fact that non-bitter mutant has never been occurred, or is due to deliberate preference of bitter types by humans for several past generations because of its typical flavor. Two classes of bitter compounds are present: an alkaloid, momordicin (Jeffrey 1980), and the triterpene glycosides, momordicoside K and L (Okabe et al. 1982). They have described these momordicosides as cucurbitacins, while Neuwinger (1994) pointed out that these compounds lack the oxygen function at C-II that characterizes 'true' cucurbitacins, the typical bitter compounds of the Cucurbitaceae, and the most bitter compounds in the plant kingdom (Johns 1990). It is possible that humans did not select for a non-bitter lineage because its fruit is consumed for its 'cooling' properties that restore the body's balance, rather than its caloric or nutritional value. In Indian Ayurvedic medicine, 'cooling' properties are attributed to many bitter substances, including *M. charantia* (Dash 1991). In India, efforts have been made to discover methods to retain the bitter flavor of *M. charantia* fruit that are preserved for consumption in the off-season (Kumar et al. 1991). Likewise, the 'cooling' effect is an important reason for its consumption in China (Duke and Ayensu 1985), where slightly bitter, medium to strongly bitter, or extremely bitter cultivars are consumed in different parts of the country (Yang and Walters 1992).

Bitter gourd is well known for its medicinal properties since ancient times. The fresh fruits have high nutritive value and can be compared with any other vegetables. The fruits are used as a tonic, antidiabetic, purgative, stomachic, anti-inflammatory, carminative, anthelminthic, febrifuge, stimulant, thermogenic, etc. The fruit contains 2.1 g of protein, 1.8 mg of iron, 20 mg of calcium, 88 mg of vitamin C, 55 mg of phosphorus, and 210 IU of vitamin A in 100 g of edible portion (Behera et al. 2007). Nearly all parts of the plants are edible; the immature fruits are boiled, curried, pickled, canned, and dehydrated. The young leaves are cooked as the vegetable in some part of India, and in the Philippines leaves are used for flavoring variety of dishes. A special type of tea is also prepared in Japan by using extracts from leaves and fruits (Reyes et al. 1994). Immature fruits are most commonly consumed plant part throughout the world. The nutritive values of the fruits vary greatly among different genotypes and accessions of this species (Behera et al. 2008a; Tan et al. 2014). Immature and dark green-colored fruits have intense bitterness than physiologically matured and lighter-colored fruits (Aminah and Anna 2011). The bitterness of fruit is associated with saponin content, mainly four cucurbitane glycosides: momordicines I and II, and momordicosides K and L (Jeffrey 1980; Okabe et al. 1982; Yasuda et al. 1984).

Bitter gourd has been studied as a complementary drug in the treatment of diabetes to reduce both glucose level and oxidative stress. It is also used as an antiviral therapy for HIV infection and as a cytostatic in certain cancers (Lee-Huang et al. 1995). During the past decade, the antidiabetic properties of the crop have been studied extensively and a hypoglycemic principle called 'charantin' has been isolated (Raman and Lau 1996; Sarkar et al. 1996). *Momordica* species are used for the treatment of both type I and II diabetes because of its several proteins

(insulin-like peptides, charantin, vicine, glycosides, karavilosides) involved in the insulin signaling pathways (Blum et al. 2012). Hazarika et al. (2012) studied the antidiabetic mechanism of *Momordica* species which involved in blocking the active site of the glycogen synthase kinase-3 (GSK-3) protein with three known antidiabetic compounds, namely charantin, momordenol, and momordicin.

There are many significant evidences suggesting links between diets rich in bitter melon and lower risk of lymphoid leukemia, lymphoma, choriocarcinoma, melanoma, breast cancer, skin tumor, prostatic cancer, squamous carcinoma of the tongue and larynx, human bladder carcinomas, and Hodgkin's disease (Basch et al. 2003). Leaf extract of bitter gourd significantly decreases the gene expression of MMP-2 and MMP-9 in breast cancer cell line (Pitchakarn et al. 2010). Most attention to date has focused on several phytoconstituents of *Momordica* including triterpenoids, Kuguacin J, α and β-momorcharin, and MAP 30 protein. MAP 30 protein isolated from seeds of bitter gourd shows inhibitory properties against growth of Hep G2 leukemia cells (Fang et al. 2012). α-momorcharin and β-momorcharin from mature seeds effectively inhibit the growth of prostate cancer (Xiong et al. 2009).

Studies on the extent of diversity available for different nutritional traits are mostly restricted to the concentration of various minerals and vitamins except a few recent studies regarding variability for important phytomedicinal traits. Jaiswal et al. (1990) conducted divergence study of different minerals and vitamins in seven genotypes of bitter gourd (Priya, Arka Harit, Sakaldiha, Jakhani-12, Jakhani-30, Pusa Do Mausami, and Baramasi). They reported significant variation for protein, carbohydrates, sugars, ash, P, Fe, total S, vitamin C, acidity, and total phenols in these cultivars. Genetic variability in ascorbic acid and total carotenoid content in Indian bitter gourd were analyzed by Dey et al. (2005) using 38 accessions. They observed significant differences among the genotypes. The range of ascorbic acid content varied from 60.20 mg to 122.07 mg/100 g of fresh weight.

Similarly, total carotenoid content ranged from 0.205 mg to 3.2 mg/100 g of fresh weight. The highest amount of ascorbic acid was found in DBTG-3 (122.07 mg), whereas DBTG-8 recorded the maximum total carotenoid content (3.2 mg). The phenolic contents and antioxidant properties of leaf, stem, and fruit extracts through 1,1-diphenyl-2-picrylhydrazyl (DPPH) were analyzed by Kubola and Siriamornpun (2008). The results revealed that the highest value of antioxidant activity, based on DPPH radical-scavenging activity and ferric reducing antioxidant power (FRAP), was recorded in leaf extract, while the green fruit extract showed the highest value of antioxidant activity, based on hydroxyl radical-scavenging activity, β-carotene–linoleate bleaching assay, and total antioxidant capacity. The predominant phenolic compounds in the fruits were gallic acid.

Concentration of essential mineral elements in bitter gourd fruit was also analyzed by Kosanovic et al. (2009), and they concluded that *M. charantia* is a good source of essentials particularly zinc, copper, molybdenum, chromium, and vanadium. Islam et al. (2011) found significant variation in phenolics contents in flesh, seed, and seed coat tissue (SCT) in four bitter gourd varieties. The four varieties were similar in their antioxidant activities of the flesh tissues; however, they were significantly different in their antioxidant activities in the seed and seed coat tissues. Ten cultivars of bitter gourd from different sources (5 from Japan and 5 from the Philippines) were studied for charantin concentration by Kyoung Kim et al. (2013). They concluded that charantin levels varied widely among the cultivars and the Japanese cultivars contained higher charantin levels than the Philippine originated genotypes. The highest content was found in the genotype Peacock from Japan and the lowest in the cultivar Trident 357 from the Philippines. The antioxidant properties of whole fruits and fruits without seeds and pith of two varieties of *M. charantia* were investigated by Choo et al. (2014). The results showed that the fruits without seeds and pith and whole fruits of the two varieties of bitter gourd exhibited different antioxidant contents and activities. The

ascorbic acid content ranged from 8.12 mg/100 g to 24.46 mg/100 g, whereas the total phenolic content ranged from 1.47 mg GAE/100 g to 27.23 mg GAE/100 g. The antiradical power ranged from 4.67 to 5.94, and the ferrous ion chelating activity using the fruit extract concentration of 0.34 g/mL ranged from 10.6 to 89.3%. Krishnendu and Nandini (2016) assessed the chemical and nutritional composition of the selected bitter gourd types (four types of commercially cultivated bitter gourd, viz., light green small, light green big, dark green small, and dark green big along with nei paval) in the fresh and dried form. The study revealed that highest protein, moisture, vitamin C, and folic acid content were found in light green big types (2.06 g, 90.40%, 98.2 mg, and 0.10 mg/ml, respectively). The highest carbohydrate and fiber content were found in the light green small types (8.22 and 1.21 g). A significant difference in β-carotene concentration of bitter gourd types was reported, and the amount of β-carotene was found to be the highest in nei paval sample (140.03 mg/100 g). In the case of mineral analysis, highest calcium, phosphorus, and sodium contents were found in light green big types (25.44 mg/l00g and 79.64 mg/100 g 20.12 mg/100 g, respectively). The potassium and iron contents were found the highest in nei paval (174.46 mg/100 g 2.14 mg), and the highest manganese, copper, and zinc contents were noticed in light green big types (34.57 mg, 40.17 mg, and 90.41 mg/l00g, respectively). Difference in concentration of the important phytochemicals like phenolic acids and carotenoids at different stages of maturation of fruits was studied by Lee et al. (2017). The study revealed that with increasing maturation, gallic acid, chlorogenic acid, and catechin contents increased, while caffeic acid, *p*-coumaric acid, and ferulic acid concentration decreased. Carotenoids' concentration increased along with increasing maturation of the fruits. They also recorded increasing activities of free radical-scavenging compounds with maturation of the fruits.

Kole et al. (2010) investigated the concentration of two important phytomedicines, cucurbitacin-B (CCR-B) and charantin (CHR) in

22 genotypes from *muricata* and *charantia* botanical varieties. They recorded a very wide range in concentration of these two phytomedicinal traits across the germplasm. CCR-B content ranged from 0.3 to 1.0 mg/g of lyophilized fruit powder with an average of 0.521 mg/g, while CHR content ranged from 0.65 to 1.35 mg/g with an average of 0.882 mg/g. Relatively higher values for the contents of CCR-B and CHR were observed, in general, in the *muricata* genotypes. Two genotypes (CBM10 and CBM12) with high concentration of CCR-B and CHR were identified for their use in the future breeding program. One advanced breeding line, CBM18, was developed with phytomedicine contents with averages of 0.817 mg/g for CCR-B and of 1.833 mg/g for CHR. They concluded that selection of genotypes belonging to the two botanical varieties as parents in crossing could provide useful segregating population for mapping genes or QTLs controlling the contents of the two phytomedicines.

4.3.3 Molecular Marker Diversity

4.3.3.1 Genetic Diversity Based on Molecular Markers

Evaluation of genetic diversity based on morphological characters alone does not reflect the extent of divergence available among the genotypes as most of the economically important traits are highly influenced by environmental factors and developmental stages of the plants. On the other side, DNA-based markers can reduce the breeding cycle and are independent of environmental conditions, which provide a quick and reliable method for estimating genetic relatedness among plant genotypes (Thormann et al. 1994), detectable in all stages of plant growth and development (Mondini et al. 2009), and show a higher level of polymorphism (Kole and Gupta 2004).

The DNA-based markers are extremely useful in developing genotypic specific fingerprints with application in germplasm characterization, hybridity testing, and cultivar identification. Till the end of twentieth century, the molecular markers used in studying genetic diversity were of three major types: viz., (i) hybridization-based markers like restriction fragment length polymorphism (RFLP); (ii) polymerase chain reaction (PCR)-based markers utilizing short random sequence such as random amplified polymorphic DNA (RAPD), arbitrary primed polymerase chain reaction (AP-PCR), DNA amplification fingerprinting (DAF), and amplified fragment length polymorphism (AFLP); (iii) PCR-based sequence-dependent markers such as sequence tagged site (STS), sequence characterized amplified regions (SCARs), and simple sequence repeats (SSRs) or microsatellites (Staub et al. 2000). In the last decade, lots of sequence-based markers including single-nucleotide polymorphisms (SNPs) and diversity array technology (DArT) were developed and used widely in studying genetic diversity with rapid advancement in next-generation sequencing (NGS) technology.

4.3.3.2 Genetic Diversity Based on RAPD, AFLP, and ISSR Markers

Studies on genetic divergence based on molecular markers in *M. charantia* were started in the last decade. In one of the earliest reports, Dey et al. (2006) conducted a study on molecular diversity in 38 bitter gourd lines using RAPD markers. Out of 116 random decamer primers screened, 29 were polymorphic and informative enough to analyze these genotypes. A total of 208 markers were generated of which 76 (36.50%) were polymorphic, and the number of bands per primer was 7.17 out of which 2.62 were polymorphic. Pair-wise genetic distance (GD) based on molecular analysis ranged from 0.07 to 0.50. However, they found that clustering pattern based on agronomic traits and molecular markers was different. In the same population, 15 inter-simple sequence repeat (ISSR) primers were used to reveal the extent of diversity which produced a total of 125 markers, 94 (74.8%) of which were polymorphic (Singh et al. 2007). The number of polymorphic markers ranged from 0 (UBC 841) to 12 (UBC 890) with a mean of 6.27 markers per primer. Pair-wise genetic distances (GDs) of the 38 bitter gourd accessions, based on the 125 markers, ranged from 0.093 (Pusa Do

Mausami—green vs. DBTG 7) to 0.516 (Pusa Do Mausami—white vs. DBTG 101). Behera et al. (2008a) reported a range of 3–15 RAPD amplicons per RAPD primer with an average of 2.6 amplicons per primer in bitter gourd. The size of amplicons varies from 200 bp to 3000 bp in a large number of bitter gourd genotypes. RAPD markers were used by Rathod et al. (2008) for assessing genetic relationships among 20 genotypes of bitter gourd. A total of 143 polymorphic amplified products were observed from 14 decamer primers, which discriminated all the accessions with a mean of 10.2 amplified bands per primer, 48.3% (69 bands) of which were polymorphic bands. A dendrogram grouped the genotypes into two clusters 'A' and 'B' at 29 linkage distances. Cluster 'A' consisted of one variety 'Arka Harit' characterized by plants that are highly susceptible to fruit fly and downy mildew infestations. Cluster 'B' was subdivided into 'B1' and 'B2' clusters at a linkage distance of 26 with one ('Nanjangood Local') and 18 genotypes, respectively. The genetic dissimilarity matrix based on squared Euclidean distance showed a maximum dissimilarity (52%) between the genotypes 'Nanjangood Local' and 'IC-42261' and a minimum dissimilarity (9%) between the genotypes 'IC-42261' and 'VRBT-93', 'DARA-1', and 'IC-065782'. In another study, Behera et al. (2008b) analyzed genetic relationships among 38 bitter gourd accessions with the aid of 29 RAPD, 15 ISSR, and six AFLP markers. More polymorphism was detected by AFLPs when compared with RAPD- and ISSR-based markers using the same germplasm array (RAPD 36.5%, ISSR 74.5%, and AFLP 78.5% polymorphism). The average marker index (MI) values derived from the three different marker systems differed dramatically, indicating that they vary in their discriminatory power (AFLP > ISSR > RAPD). The AFLP markers used were only weakly correlated with ISSR ($r^2 = 0.007$) and RAPD ($r^2 = 0.04$) marker data, whereas a comparatively high correlation ($r^2 = 0.77$) was found between RAPD and ISSR marker systems. Genetic diversity of bitter gourd (*Momordica charantia* L.) based on RAPD and ISSR markers was evaluated by Gao et al.

(2010). The results showed that 93 and 81 bands were obtained by RAPD and ISSR markers amplified through 10 selected primers, respectively. The percentage of polymorphic bands (PPB) in ISSR detection of 61.29% was higher than that in RAPD with 50.54%. The germplasm was divided into three main clusters with six subgroups using RAPDs. This classification based on RAPDs was similar to the phenotype classification of bitter gourd. They were also divided into three main clusters and seven subgroups by ISSR which was consistent with the color classification of bitter gourd. Using RAPD and ISSR markers, Dalamu et al. (2012) assessed genotypic variation among 50 indigenous and exotic bitter gourd genotypes. The results revealed that out of 84 reproducible amplicons generated by 17 RAPD primers, 33 were polymorphic and out of 58 reproducible amplicons generated by 11 ISSR primers, 41 were polymorphic. The polymorphic information content (PIC), resolving power (RP), and marker index (MI) were 0.17, 1.14, and 0.82, respectively, for RAPD markers, whereas ISSR markers showed comparatively high polymorphic information content (0.40), resolving power (1.87), and marker index (2.11).

4.3.3.3 Genetic Diversity Based on SSR Markers

Among the DNA markers, simple sequence repeats (SSRs) are an ideal genetic marker type and have gained significant importance in plant genetics and breeding owing to its desirable attributes like multi-allelic nature, codominant inheritance (Akkaya et al. 1992). These also exhibit extensive genome coverage and require small amount of starting DNA, ease of detection by PCR, and capillary electrophoresis (Morgante and Olivieri 1992; Peakall et al. 1998), characterized by high polymorphism (Gupta et al. 1996). However, only a few microsatellite markers are available in bitter gourd and 16 SSRs were developed using fast isolation by AFLP of sequences containing repeats (FIASCO) technique (Wang et al. 2010; Guo et al. 2012), 11 through genomic library enrichment (Ji et al. 2012) and 43 from other cucurbits through cross-species transferability (Chiba et al. 2003; Watcharawongpaiboon and Chunwongse 2008; Xu et al. 2011). Polymorphic microsatellite markers were developed by Wang et al. (2010) for *M. charantia* to investigate the genetic diversity and population structure within and between *M. charantia* and its four related species (*Cucurbita pepo* L., *Luffa cylindrica* L., *Lagenaria siceraria* L., and *Cucumis sativus* L.). Using FIASCO method, 16 polymorphic microsatellite loci were identified in 36 individuals of *M. charantia*. Across all the *M. charantia* samples, the number of alleles per locus ranged from three to eight. Seven primers successfully amplified in the four related species. Guo et al. (2012) also used FIASCO method to develop and characterize 25 microsatellite markers from the genome of *M. charantia* for their application in studying diversity and construction of population structure. Their findings showed that 10 loci were polymorphic, and the number of alleles per locus ranged from 3 to 7, with the observed heterozygosity ranging from 0.46 to 0.65. These developed markers also amplified successfully in the related species *M. cochinchinensis* and *Cucurbita pepo*. Guang-guang et al. (2013) studied genetic diversity and relationship among 50 bitter gourd varieties using 16 pairs of SSR primers. They were able to amplify 90 alleles using these 16 SSRs, and the average values of polymorphism loci percentage (P), genetic heterozygosity (He), effective number of alleles (Ne), Shannon–Wiener index (H'), PIC, and genetic similarity coefficient were 97.78, 0.853, 7.728, 2.087, 0.834, and 0.706%, respectively. The genotypes were classified into 6 groups using UPGMA methods. One hundred sixty novel microsatellite markers in *M. charantia* through sequencing of small insert genomic library were developed by Saxena et al. (2014). Genetic diversity in 54 bitter gourd accessions was studied using the developed SSRs, and it was found that 20% of the loci were polymorphic with the PIC values ranging from 0.13 to 0.77. Fifteen Indian varieties were clearly distinguished indicative of the usefulness of the developed markers. These markers are highly transferable to six other

Momordica species and may be used as an efficient tool in phylogenetic and comparative studies among the *Momordica* species. Dhillon and Sanguansil (2016) studied genetic diversity among a large collection of bitter gourd from different parts of Asia (114 accessions) which comprises several landraces, breeding lines, and commercial cultivars. Neighbor-joining tree analysis revealed a high level of genetic variability in the collection. The 114 accessions formed three subpopulations represented by five clusters. Distribution of accessions across the five clusters reflected their geographic origin to a large extent. South Asian accessions originating from India, Bangladesh, and Pakistan were more closely related to each other than to any other geographical group.

4.3.4 Population Structure Study

In bitter gourd, only a limited number of studies were conducted to classify the genotypes based on population structure analysis. Cui et al. (2018) identified a total of 188,091 and 167,160 SSR motifs in the genomes of the bitter gourd lines 'Dali-11' and 'OHB3-1,' respectively, based on the draft genome sequence. Subsequently, the SSR content, motif lengths, and classified motif types were characterized for the bitter gourd genomes and compared among all the cucurbit genomes. Lastly, they had designed a large set of 138,727 unique in silico SSR primer pairs. Among them, 71 primers were successfully amplified in two bitter gourd lines 'Dali-11' and 'K44'. They have used 21 SSR markers to analyze a collection of 211 bitter gourd lines from all over the world. STRUCTURE was used to infer population structure of the bitter gourd samples (for $K = 2, 3, …, 10$). At $K = 3$, the clustering of samples was the most appropriate because it produced the highest 1 K value (373.87). Accordingly, the total panel was divided into three main populations that were labeled P1, P2, and P3, consisting of 23, 85, and 103 samples, respectively. This result was consistent with the NJ tree and geographic origins of the samples. In a collection of 22 genotypes from *charantia* and *muricata* groups, population structure analysis was attempted by Kole et al. (2010). An admixture model for population structure analysis revealed the population to be optimally stratified into four subpopulations with fixation indices values of 0.785, 0.663, 0.542, and 0.582 for the four subpopulations with the highest diversity in subpopulation-1. Each of the subpopulations was represented by both *charantia* and *muricata* genotypes. However, there was no correlation between population structure of the genotypes and their geographical origin. The lack of correlation between population structure and geographical origin could be due to the fact that the country of domestication might not necessarily represent the actual origin of the genotypes.

4.4 Conclusion and Future Perspective

Based on the reports on diversity studies, it is evident that substantial diversity is available in *Momordica* spp. for different economically important traits. However, availability of diversity in terms of sex expression is limited to a few genotypes for India, China, and Japan. There is an urgent need to identify gynoecious lines in different groups of bitter gourd for their successful utilization in hybrid breeding program. Moreover, there is very limited number of studies available regarding the extent of divergence available in bitter gourd genotypes for important nutritional traits. It is necessary to identify genotypes with higher concentration of charantin, momordicin, saponin, and insulin-like compounds. Availability of genotypes from different groups with higher concentration of important phytochemicals will help in developing breeding lines with better nutritional values. Moreover, it is necessary to identify genotypes with adaptation to different environments and cultivation conditions. Identification of gynoecious parthenocarpic genotypes has potential to revolutionize the bitter gourd cultivation under protected condition. It is important to take research works on mutations for induction of desirable traits in the available genotypes and create more genetic diversity. Induction of haploids through gynogenesis, pollination with irradiated pollen grains, and androgenesis could help in creating more diversity in bitter gourd population. Studies on population structure and stratification of genotypes to understand the actual diversity available in the *Momordica* population are very limited. Therefore, it is needed to undertake the analysis of population structure based on the available draft genome sequence using sequence-based markers.

References

Agasimani SC, Salimath PM, Dharmatti PR (2008) Stability for fruit yield and its components in bitter gourd (*Momordica charantia* L.). Veg Sci 35(2):140–143

Akkaya MS, Bhagwat AA, Cregan PB (1992) Length polymorphisms of simple sequence repeat DNA in soybean. Genetics 132(4):1131–1139

Ames O (1939) Economic annuals and human cultures. Botanical Museum of Harvard University, Cambridge

Aminah A, Anna PK (2011) Influence of ripening stages on physicochemical characteristics and antioxidant properties of bitter gourd (*Momordica charantia* L.). Intl Food Res J 18:895–900

Basch E, Garbardi S, Ulbricht C (2003) Bitter melon (*Momordica charantia*): a review of efficacy and safety. Am J Health-Syst Pharm 60:356–359

Behera TK, Dey SS, Sirohi PS (2006) DBGy-201 and DBGy-202: two gynoecious lines in bitter gourd (*Momordica charantia* L.) isolated from indigenous source. Indian J Genet 66:61–62

Behera TK, Dey SS, Munshi AD, Gaikwad AB, Pal A, Singh I (2009) Sex inheritance and development of gynoecious hybrids in bitter gourd (*Momordica charantia* L.). Sci Hort 120:130–133

Behera TK, Gaikwad AB, Singh AK, Staub JE (2008a) Relative efficiency of DNA markers (RAPD, ISSR and AFLP) in detecting genetic diversity of bitter gourd (*Momordica charantia* L.). J Sci Food Agri 88(4):733–737

Behera TK, Singh AK, Staub JE (2008b) Comparative analysis of genetic diversity in Indian bitter gourd (*Momordica charantia* L.) using RAPD and ISSR 146 markers for developing crop improvement strategies. Sci Hort 115:209–217

Behera TK, Staub JE, Behera S, Simon PW (2007) Bitter gourd and human health. Med Aromat Plant Sci Biotechnol 1:224–226

Bharathi LK (2010) Phylogenetic Studies in Indian *Momordica* Species, Ph.D. Thesis. Division of Vegetable Science, IARI, New Delhi, India

Bhave SG, Mehta JL, Bendale VW, Mhatre PP, Pethe UB (2003) Character association and path coefficient analysis of bitter gourd (*Momordica charantia* L.). Orisaa J Hort 31(1):44–46

Blum A, Loerz C, Martin HJ, Claudia A, Staab-Weijnitz Maser E (2012) *Momordica charantia* extract, a herbal remedy for type-2 diabetes, contains a specific 11β-hydroxysteroid dehydrogenase type 1 inhibitor. J Steroid Biochem Mol Biol 128:51–55. https://doi.org/10.1016/j.jsbmb.2011.09.003

Burkhill IH (1935) A dictionary of the economic products of the malay penninsula. Crown Agents for the Colonies, London

Chakravarty HL (1959) Monograph of Indian Cucurbitaceae. Rec Bot Surv India 17:81

Chiba N, Suwabe K, Nunome T, Hirai M (2003) Development of microsatellite markers in melon (*Cucumis melo* L.) and their application to major Cucurbit crops. Breed Sci 53:21–27

Choo WS, Yap JY, Chan SY (2014) Antioxidant properties of two varieties of bitter gourd (*Momordica charantia*) and the effect of blanching and boiling on them. J Trop Agri Sci 37(1):121–131

Coe FG, Anderson GJ (1996) Ethnobotany of the Garifuna of Eastern Nicaragua. Econ Bot 50(1):71–107

Cui J, Luo S, Niu Y, Huang R, Wen Q, Su J et al (2018) A RAD-based genetic map for anchoring scaffold sequences and identifying QTLs in bitter gourd (*Momordica charantia*). Front Plant Sci 9:477. https://doi.org/10.3389/fpls.2018.00477

Dalamu Behera TK, Gaikwad AB, Saxena S, Bharadwaj C, Munshi AD (2012) Morphological and molecular analyses define the genetic diversity of Asian bitter gourd (*Momordica charantia* L.). Austral J Crop Sci 6(2):261–267

Dash B (1991) Materia medica of ayurveda based on Madanapala's Nighantu. B. Jain Publishers, New Delhi, India

De Wilde WJJO, Duyfjes BEE (2002) Synopsis of *Momordica* (Cucurbitaceae) in South East Asia and Malaysia. Bot Zhurn 57:132–148

Decker-Waiters DS (1999) Cucurbits, sanskrit, and the Indo-Aryans. Econ Bot 53(1):98–112

Dey SS, Behera TK, Munshi AD, Bhatia R (2009) Genetic variability, genetic advance and heritability in bitter gourd (*Momordica charantia* L.). Indian Agri 53 (1/2):7–12

Dey SS, Behera TK, Pal A, Munshi AD (2005) Correlation and path coefficient analysis in bitter gourd (*Momordica charantia* L.). Veg Sci 32(2):173–176

Dey SS, Singh AK, Chandel D, Behera TK (2006) Genetic diversity of bitter gourd (*Momordica charantia* L.) genotypes revealed by RAPD markers and agronomic traits. Sci Hort 109:21–28

Dhillon NPS, Sanguansil S (2016) Diversity among a wide Asian collection of bitter gourd landraces and their genetic relationships with commercial hybrid cultivars. J Amer Soc Hort Sci 141(5):475–484

Duke JA, Ayensu ES (1985) Medicinal plants of China, vol 1. Reference Publications, Algonac, MI

Fang EF, Zhang CZY, Wong JH, Shen JY, Li CH, Bun NGT (2012) The MAP30 protein from bitter gourd (*Momordica charantia*) seeds promotes apoptosis in liver cancer cells in vitro and in vivo. Cancer Lett 324:66–74

Gao S, Lin BY, Xu DX, Fu RQ, Lin F, Lin Y-Z, Pan DM (2010) Genetic diversity of bitter gourd (*Momordica charantia* L.) based on RAPD and ISSR. J Plant Genet Resour 11:78–83

Guang-guang LI, Yan-song Z, Xiang-yang LI, Zhang H, Pei-guo G, Hong-di H (2013) Genetic diversity analysis in bitter gourd germplasm resources based on SSR molecular markers. J Southern Agri 2013-01

Guo D, Zhang J, Xue Y, Hou X (2012) Isolation and chararacterization of 10 SSR markers of *Momordica charantia* (*Cucurbitaceae*). Amer J Bot 97:e182–e183

Gupta PK, Balyan HS, Sharma PC, Ramesh B (1996) Microsatellites in plants: a new class of molecular markers. Curr Sci 70:45–54

Harlan JR, Wet JMJD (1971) Toward a rational classification of cultivated plants. Taxon 20(4):509–517

Hazarika R, Parida P, Neog B, Yadav RNS (2012) Binding energy calculation of GSK3 protein of human against some anti-diabetic compounds of *Momordica charantia* Linn (Bitter melon). Bioinformation 8:251–254

Iqbal M, Munawar M, Najeebullah M, Ahmad D (2016) Assessment of genetic diversity in bitter gourd. Intl J Veg Sci 22(6):578–584

Islam S, Jalaluddin M, Hettiarachchy NS (2011) Bioactive compounds of bitter melon genotypes (*Momordica charantia* L.) in relation to their physiological functions. Funct Food Health Dis 2:61–74

Iwamoto E, Ishida T (2006) Development of gynoecious inbred line in balsam pear (*Momordica charantia* L.). Hort Res (Japan) 5:101–104

Jaiswal RC, Kumar S, Singh D K and Kumar S (1990) Variation in quality traits in bitter gourd. Veg Sci 17: 186–190

Jeffrey C (1980) A review of the Cucurbitaceae. Bot J Linn Soc 81(3):233–247

Jeffrey C (2001) Cucurbitaceae. In: Hanelt P (ed) Encyclopedia of agricultural and horticultural crops, vol 3. Springer, Berlin, pp 1510–1557

Ji Y, Luo Y, Hou B, Wang W, Zhao J, Yang L, Xue Q, Ding X (2012) Development of polymorphic microsatellite loci in *Momordica charantia* (Cucurbitaceae) and their transferability to other cucurbit species. Sci Hort 140:115–118

Johns T (1990) With Bitter Herbs They Shall Eat It. University of Arizona Press, Tucson

Jongkind CCH (2002) A new species of *Momordica* (Cucurbitaceae) from West Africa. Blumea 47:343–345

Joseph JK (2005) Studies on ecogeography and genetic diversity of the genus Momordica L. in India. Dissertation, Mahatma Gandhi University, Kottayam, Kerala

Joseph JK, Antony VT (2007) *Momordica sahyadrica*sp. nov. (Cucucrbitaceae), an endmic species of Western Ghats of India. Nord J Bot 24:539–542

Khajuria S, Thomas J (1993) Traditional Indian beliefs about the dietary management of diabetes: an exploratory study of the implications for the management of Gujarati diabetics in Britain. J Hum Nutr Diet 5(5):311–321

Khan MH, Bhuiyan KC, Saha MR, Ali SMY (2015) Variability, correlation and path co-efficient analysis of bitter gourd (*Momordica charantia* L.). Bangladesh J Agri Res 40(4):607–618

Kole C, Gupta PK (2004) Genome mapping and map based cloning. In: Jain HK, Kharkwal MC (eds) Plant breeding—from mendelian to molecular approaches. Narosa Publishing House, New Delhi, India, pp 255–299

Kole C, Olukolu B, Kole P, Abbott AG (2010) Association mapping of fruit traits and phytomedicine contents in a structured population of bitter melon (*Momordicacharantia* L.). Cucurbitaceae 2010, Nov 14–18 (2010), South Carolina, USA

Kosanovic M, Hasan MY, Petroianu G, Marzouqi A, Abdularhman O, Adem A (2009) Assessment of essential and toxic mineral elements in bitter gourd (*Momordica charantia*) fruit. Intl J Food Prop 12(4):766–773

Krishnendu JR, Nandini PV (2016) Nutritional composition of bitter gourd types (*Momordica Charantia* L.). Intl J Adv Eng Res Sci 3(10):096–104

Kubola J, Siriamornpun S (2008) Phenolic contents and antioxidant activities of bitter gourd (*Momordica charantia* L.) leaf, stem, and fruit fraction extracts in vitro. Food Chem 110:881–890

Kumar SS, Kalra R, Nath N (2008) Dehydration of bitter gourds (*Momordica charantia* Linn) rings. J Food Sci Technol 28(1):52–53

Kyoung Kim Y, Park W, Uddin MR, Bok Kim Y, Bae H, Kim H, Woong Park K, Un Park S (2013) Variation of charantin content in different bitter melon cultivars. Asian J Chem 26:309–310

Lee SH, Jeong YS, Song J, Hwang Kyung-A, Noh GM, Hwang IG (2017) Phenolic acid, carotenoid composition, and antioxidant activity of bitter melon (*Momordica charantia* L.) at different maturation stages. Intl J Food Prop 20:S3078–S3087

Lee-Huang S, Huang PL, Chen HC, Bourinbaiar A, Huang HI, Kuang HF (1995) Anti-HIV and anti-tumor activities of recombinant MAP30 from bitter melon. Gene 161:151–156

Lira R, Caballero J (2002) Ethnobotany of the wild Mexican Cucurbitaceae. Econ Bot 56(4):380–398

Matsumura H, Miyag N, Taniai N, Fukushima M, Tarora K, Shudo A et al (2014) Mapping of the gynoecy in bitter gourd (*Momordica charantia*) using RAD-Seq analysis. PLoS ONE 9:e87138

Miniraj N, Prasanna KP, Peter KV (1993) Bitter gourd (*Momordica* spp.) In: Kalloo G, Bergh BO (eds) genetic improvement of vegetable crops. Pergamon Press, Oxford, pp 239–246

Mondini L, Noorani A, Pagnotta MA (2009) Assessing plant genetic diversity by molecular tools. Diversity 1:19–35

Morgante M, Olivieri A (1992) PCR-amplified microsatellites as markers in plant genetics. Plant J 3:175–182

Neuwinger HD (1994) African ethnobotany, poisons and drugs. Chapman and Hall, London

Ng TB, Chan WY, Yeung HW (1992) Proteins with abortifacient, ribosome inactivating, immunomodulatory, antitumor and anti-AIDs activities from Cucurbitaceae plants. Gen Pharmacol 23:579–590

Okabe H, Miyahara Y, Yamauchi T (1982) Studies on the constituents of *Momordica charantia* L. III. Chem Pharm Bull 30(11):3977–3986

Peakall R, Gilmore S, Keys W, Morgante M, Rafalski A (1998) Cross-species amplification of Soybean (*Glycine max*) simple sequence repeats (SSRs) within the genus and other legume genera: Implications for the transferability of SSRs in plants. Mol Boil Evol 15:1275–1287

Perry LM (1980) Medicinal plants of East and Southeast Asia. MIT Press, Cambridge, MA

Pitchakarn P, Ogawa K, Suzuki S, Takahashi S, Asamoto M, Chewonarin T, Limtrakul P, Shirai T (2010) *Momordica charantia* leaf extract suppresses rat prostate cancer progression in vitro and in vivo. Cancer Sci 101:2234–2240

Platel K, Srinivasan K (1995) Effect of dietary intake of freeze dried bitter gourd (*Momordica charantia*) in streptozotocin induced diabetic rats. Narhung 39(4):3977–3986

Ram D, Kumar S, Banerjee MK, Kalloo G (2002) Occurrence, identification and preliminary characterization of gynoecism in bitter gourd. Indian J Agri Sci 72(6):348–349

Ram D, Kumar S, Singh M, Rai M, Kalloo G (2006) Inheritance of gynoecism in bitter gourd (*Momordica charantia* L.). J Hered 97:294–295

Raman A, Lau C (1996) Anti-diabetic properties and phytochemistry of Momordica charantia L. (Cucurbitaceae). Phytomedicine 2:349–362

Rathod V, Narasegowda NC, Papanna N, Simon L (2008) Evaluation of genetic diversity and genome fingerprinting of bitter gourd genotypes (*Momordica charantia* L.) by morphological and RAPD markers. Intl J plant Breed 2:79–84

Reyes MEC, Gildemacher BH, Jansen GJ (1994) *Momordica charantia*L. In: Siemonsma JS, Piluek K (eds) Plant resources of South-East Asia: vegetables. Pudoc Scientific Publishers, Wageningen, The Netherlands, pp 206–210

Robinson HF (1966) Quantitative genetics in relation to breeding on centennial of mendelium. Indian J Genet 26:171–187

Robinson RW, Decker-Walter DS (1999) Cucurbits. CAB International, Wallingford, Oxon, U.K

Sahni GP, Singh RK, Saha BC (1987) Genotypic and phenotypic variability in ridge gourd. Indian J Agri Res 57:666–668

Sarkar S, Pravana M, Marita R (1996) Demonstration of the hypoglycemic action of *Momordica charantia* in a validated animal model of diabetes. Pharma Res 33(1):1–4

Saxena S, Singh A, Archak S, Behera TK, John JK, Meshram SU, Gaikwad AB (2014) Development of novel simple sequence repeat markers in bitter gourd (*Mordica charantia* L.) through enriched genomic libraries and their utilization in analysis of genetic diversity and cross-species transferability. Appl Biochem Biotechnol https://doi.org/10.1007/s12010014-1249-8

Singh AK, Behera TK, Chandel D, Sharma P, Singh NK (2007) Assessing genetic relationships among bitter gourd (*Momordica charantia* L.) accessions using inter-simple sequence repeat (ISSR) markers. J Hort Sci Biotechnol 82 (2): 217–222

Singh AK (1990) Cytogenetics and evolution in the cucurbitaceae. In: Bates DM, Robinson RW, Jeffrey C (eds) Biology and utilization of Cucurbitaceae. Comstock Publishing Associates, Cornell University Press, Ithaca, New York and London, pp 10–28

Singh V, Rana DK, Shah KN (2017) Genetic variability, heritability and genetic advance in some strains of bitter gourd (*Momordica charantia* L.) under subtropical conditions of Garhwal Himalaya. Plant Arch 17 (1):564–568

Smith AC (1981) Flora vitiensis nova. SB Printers, Honolulu, HI

Srivastava TN (1976) Flora gorakpurensis. Today and Tomorrow Publishers, New Delhi, p 149

Staub JE, Danin-Paleg Y, Horejsi TF, Reis N, Katzir N (2000) Comparative analysis of cultivated melon groups (*Cucumis melo* L.) using RAPD and SSR markers. Euphytica 115:225–241

Tan SP, Parks SE, Stathopoulos CE, Roach PD (2014) Greenhouse-grown bitter melon: production and quality characteristics. J Sci Food Agri 94:1896–1903. https://doi.org/10.1002/jsfa.6509

Thormann CE, Camargo MEA, Osborn TC (1994) Comparison of RFLP and RAPD markers to estimating genetic relationships within and among Cruciferous species. Theor Appl Genet 80:973–980

Tiwari N, Pandey A, Singh U, Singh V (2018) Genetic variability, heritability in narrow sense and genetic advance percent of mean in bitter gourd (*Momordica charantia* L.). J Pharma Phytol 7(2):3868

Trivedi RN, Roy RP (1972) Cytological studies in some species of *Momordica*. Genetica 43:282–291

Walters TW, Decker-Walters DS (1988) Balsam-pear (*Momordica charantia*, Cucurbitaceae). Econ Bot 42 (2):286–292

Wang SZ, Pan L, Hu K, Chen C, Ding Y (2010) Development and characterization of polymorphic microsatellite markers in *Memordicacharantia* (*Cucurbitaceae*). Amer J Bot 97:e75–e78

Watcharawongpaiboon N, Chunwongse J (2008) Development and characterization of microsatellite markers from an enriched genomic library of cucumber (*Cucumis sativus*). Plant Breed 127:74–81

Whistler WA (1990) Ethnobotany of the cook Islands: the plants, their Maori names, and their uses. Allertonia 5 (4):392

Xiong SD, Yu K, Liu XH, Yin LH, Kirschenbaum A, Yao S, Narla G, DiFeo A, Wu JB, Yuan Y, Ho S, Lam YW, Levine AC (2009) Ribosome-inactivating proteins isolated from dietary bitter melon induce apoptosis and inhibit histone deacetylase-1 selectively in premalignant and malignant prostate cancer cells. Intl J Cancer 125:774–782. https://doi.org/10.1002/ijc.24325

Xu P, Wu X, Luo J, Wang B, Liu Y, Ehlers JD, Wang S, Lu Z, Li G (2011) Partial sequencing of the bottle gourd genome reveals markers useful for phylogenetic analysis and breeding. BMC Genom 12:467

Yadav M, Singh DB, Chaudhary R, Singh D (2008) Genetic variability in bitter gourd (*Momordica charantia* L.). J Hort Sci 3(1):35–38

Yang SL, Walters TW (1992) Ethnobotany and the economic role of the Cucurbitaceae in China. Econ Bot 46(4):349–367

Yasuda M, Iwamoto M, Okabe H, Yamauchi T (1984) Structures of momordicines I, II, and III, the bitter principles in the leaves and vines of *Momordica charantia* L. Chem Pharm Bull (Tokyo) 32:2044–2047

Cytogenetical Analysis of Bitter Gourd Genome

5

Ricardo A. Lombello

Abstract

There are many studies about chromosomal characteristics of *Momordica charantia*, known as bitter gourd, a vegetable species with medicinal properties. The published data include characterization of the meiosis; haploid and diploid chromosome number; description of chromosome morphology, with chromosome length and primary constriction position; more recently there has been a focus of cytomolecular investigations, with localization of ribosomal DNA (rDNA) sequences and distribution pattern of heterochromatin bands. Usually, the *M. charantia* karyotype is presented as symmetric, composed of 22 small chromosomes with similar morphology, with two pairs of 45S rDNA, colocalized with CMA positive bands, and one pair of 5S rDNA. The chromosome counts for *Momordica* genus indicate two basic numbers: $x = 11$ and $x = 14$, dividing the genus into two groups with distinct characteristics. Polyploid counts were observed in both species' groups. Variation in number and position of the cytogenetical markers were registered between the different species, evidenced by CMA fluorochrome banding. The karyotype evolution for *Momordica* may be related to the structural chromosome mutation associated with numeric chromosome variations.

5.1 Introduction

The *Momordica* L. (Cucurbitaceae) genus has been known for its horticultural use and medicinal applications, and the interest in these species justifies the cytological studies that have been carried on. This chapter aims to present the most expressive cytogenetic data published for this economically relevant genus, particularly for *M. charantia*, the most cultivated species of the genus, and indicate the direction for future research focused on the group. The cytogenetical parameters applied to taxonomy, phylogenetic systematic and management are haploid and diploid chromosome number; karyotype features, such as chromosome length and morphology; meiotic process characterization, chromosome mutation derived from irregular pairing and segregation of homologous chromosomes and respective chromatids, as well as the impact on regular and viable gametophyte production; cytomolecular characterization, including heterochromatic banding and specific gene localization. The scientific information for *Momordica* genus includes the data mentioned, essentially for species with horticultural or medicinal interest. There are more than 140 species accepted or

R. A. Lombello (✉)
Center for Natural and Human Sciences, Federal University of ABC, São Bernardo do Campo, SP, Brazil
e-mail: ricardo.lombello@ufabc.edu.br

© Springer Nature Switzerland AG 2020
C. Kole et al. (eds.), *The Bitter Gourd Genome*, Compendium of Plant Genomes,
https://doi.org/10.1007/978-3-030-15062-4_5

in study for this genus (The Plant List 2010), only 11 species were characterized cytogenetically, and most of all the studies are restricted to mention species chromosome number (Riley 1960; Mangenot and Mangenot 1962; Auquier and Renard 1975).

Some studies present more complete cytogenetical characterization (Trivedi and Roy 1972; Chattopadhyay and Sharma 1991; Ghosh et al. 2018). The published data are critically analyzed in this review, along with the impact in the knowledge of *Momordica* karyotype evolution process, breeding programs and agronomic production.

5.1.1 Chromosome Number Evolution

Most of the *Momordica* species cytogenetically studied presented chromosome diploid number $2n = 22$ (Table 5.1). This counting indicates a basic chromosome number of $x = 11$ for the genus, as proposed by Whitaker (1933). Reports of different chromosome numbers multiple from 11 also occur. Trivedi and Roy (1973) registered $2n = 33$ for *M. charantia*, and Mangenot and Mangenot (1962) described $2n = 44$ for *M. foetida*. These studies evidenced the polyploidy occurrence in different *Momordica* species. The importance of this karyotype evolution process inside a genus has been discussed previously. For Cucurbitaceae, the relative importance of polyploidy was analyzed by Agarwal and Roy (1976), Singh (1979) and Beevy and Kuriachan (1996). Besides the interspecific polyploidy variation of *Momordica*, for some species intraspecific chromosome number variations were described. *M. charantia*, *M. cymbalaria*, *M. dioica* and *M. foetida* have discrepant chromosome counts, due to polyploidy events, such as $2n = 22 = 2x$, $2n = 33 = 3x$ for *M. charantia* and $2n = 22 = 2x$, $2n = 44 = 4x$ for *M. foetida*. Diploid counting of $2n = 28$ is also presented for four *Momordica* species, *M. cochinchinensis*, *M. dioica*, *M. cardiospermoides* and *M. sahyadrica*.

Usually, as mentioned previously and pointed out by Bharathi et al. (2011), the monoecious species of *Momordica* genus presents chromosome number based on $x = 11$, and respective polyploidy variations, including $2n = 33$ for *M. charantia*. Dioecious species *M. dioica*, *M. cochinchinensis* and *M. sahyadrica* have chromosome number based on $x = 14$ (Table 5.1). This basic chromosome number had been proposed for the genus by Richharia and Ghosh (1953), Roy et al. (1966) and Trivedi and Roy (1972). The discrepant basic chromosome number drives the establishment of two groups inside *Momordica* genus, species with $x = 11$ and $x = 14$. There are also divergent chromosome basic numbers for some species; for *M. cymbalaria* there are data of haploid counting $n = 8$ (Mehetre and Thombre 1980) and diploid $2n = 18$ (Bharathi et al. 2011). The *M. dioica* related $n = 11$ (Ayyangar and Sampathkumar 1978) and $2n = 22$ (Ayyangar 1976) which are also in discrepancy from the presented $x = 14$ for this species. Bano et al. (2019) compared *M. charantia* meiotic process in natural and cultivated populations and showed different pollen mother cell (PMC) chromosome numbers, some with $n = 10$ and $n = 12$ and most with $n = 11$. The authors point out that the failure to separate homologous chromosome, or sister chromatids, is the possible cause of this numeric variation. Once the meiotic irregularities are maintained inside the populations, it can be denominated as a cytotype (Ramsey and Schemske 1998; Soltis et al. 2015). If reproductive barriers are established between the cytotypes, the speciation process takes place (De Storme and Mason 2014).

This variation in chromosome number can be understood as disploidy or aneuploidy. Disploidy variation, due to chromosomal rearrangement, such as fusions and fissions, does not involve DNA gain or loss, preserving the genome but with changes in chromosome number (Luceño and Guerra 1996). However, aneuploidy implicates one or few chromosome gain or loss, which results in distinct chromosome number and DNA content (De Storme and Mason 2014). Both events are important for the karyotype derivation, once they create reproductive barriers inside a population, based on the numeric chromosome incompatibility, possibly directing speciation.

Table 5.1 Chromosome numbers of *Momordica* species with haploid (*n*), diploid (2*n*) and authors' references, based on chromosome counts database (Rice et al. 2014)

Species	*n*	2*n*	Reference
Momordica balsamina L.	11		Trivedi and Roy (1972)
		22	Roy et al. (1966)
M. charantia L.	11		Shibata (1962)
		22	Trivedi and Roy (1972)
		33	Trivedi and Roy (1973)
M. clematidea Sond.		28	Riley (1960)
M. cochinchinensis (Lour.) Spreng.	14	28	Sen and Datta (1975)
M. cymbalaria Hook.	8		Mehetre and Thombre (1980)
	11		Beevy and Kuriachan (1996)
		18	Bharathi et al. (2011)
M. denudata C. B. Clarke	14		Beevy and Kuriachan (1996)
M. dioica Roxb. ex Willd.	11		Ayyangar and Sampathkumar (1978)
	14		Trivedi and Roy (1972)
		22	Ayyangar (1976)
		28	Roy et al. (1966)
		42	Agarwal and Roy (1976)
		56	Roy et al. (1966)
M. foetida Schumach.		22, 44	Mangenot and Mangenot (1962)
M. leiocarpa Gilg.		22	Auquier and Renard (1975)
M. sahyadrica Kattuk. and V. T. Antony		28	Barathi (2011)
M. subangulata subsp. *renigera* (G. Don) de Wilde		56	Roy et al. (1966)

The event was described for *Boechera* (Brassicaceae) genus, by Mandáková et al. (2015), with the diminishing of basic chromosome number from $x = 8$ to $x = 7$, reducing species hybridization inside the genus.

Intrageneric reproductive barriers due to chromosome number variation were also reported for *Momordica*. Trivedi and Roy (1972) performed interspecific crossing between *M. balsamina* (2*n* = 22) and *M. dioica* (2*n* = 28) and between *M. charantia* (2*n* = 22) and *M. dioica* (2*n* = 28). The chromosome number incompatibility prevented the suggested crossings, characterizing the reproductive barrier between genomes based on different basic chromosome numbers of $x = 11$ (*M. balsamina* and *M. charantia*) and $x = 14$ (*M. dioica*).

A similar reproductive incompatibility was not detected if the suggested crossings were made in populations with different ploidy levels. Trivedi and Roy (1973) crossed 2*n* = 22 *M. charantia* with tetraploid individuals, induced by treatment with the antimitotic colchicine. From these crossings, approximately 10% of fruit setting was observed. Even crossings between different *Momordica* species with the same basic chromosome number (*x*), but different chromosome counts, resulted in fruit setting and viable seed production. Mohanty et al. (1994) had high index of fruit establishment after crossing individuals of *M. dioica* (2*n* = 28) and *M. subangulata* ssp. *renigera* (2*n* = 56). Similar results were reported by Bharathi et al. (2011) in *M. dioica* and *M. subangulata* ssp. *renigera* crossings, resulting in F_1 hybrids ($2n = 42$) with low pollen viability (16.7%).

The results of the interspecific crossing enhance the existence of two groups based on basic chromosome number for *Momordica*. However, there are some data that do not correspond to these groups. Chromosome number presented for *M. cymbalaria* $n = 8$ and $2n = 18$ leads to the question about the taxonomic positioning of the species. Bharathi et al. (2011) analyzed microsporogenesis, registered nine bivalents in metaphase I in meiosis and confirmed the diploid number $2n = 18$. The authors presented a cladogram for the genus based on 27 phenotypic characteristics and cytogenetic analyses, and in this proposition *M. cymbalaria* relates distantly from the other *Momordica* species and cannot be placed in any of the groups of basic chromosome number or reproductive strategy. Beevy and Kuriachan (1996) described that the counting of $n = 8$ and $2n = 18$ for *M. cymbalaria* indicates basic chromosome number of $x = 8$ and $x = 9$ derived by decreasing aneuploidy from $x = 14$, the proposed ancestral basic chromosome number for the Cucurbitaceae family.

Cucurbitaceae family presents various basic chromosome numbers, such as $x = 7$, 8, 9, 11, 12, 13, 14, 15, 16 and 20, and the most frequent basic number related is $x = 12$ (Beevy and Kuriachan 1996). Some authors indicate that the most important processes of chromosome number variation in Cucurbitaceae family are allopolyploidy and chromosome fragmentation, resulting in the gain of chromosome number, due to the de novo synthesis of centromeric regions in the fragmented chromosomes. This occurs in different Cucurbitaceae genus, including *Momordica* (Bhaduri and Bose 1947). Segmental allopolyploidy seems to be the origin of the species *M. subangulata* subsp. *renigera* ($2n = 56$). Reproductive analyses indicate that this species is derived from the hybridization of *M. dioica* ($2n = 28$) and *M. cochinchinensis* ($2n = 28$) (Bharathi et al. 2010). Future studies based on genetic analyses can assist the correct positioning of *M. cymbalaria* in *Momordica* genus and also in the Cucurbitaceae family.

5.1.2 Karyomorphological Analysis

There is karyomorphological information for several of the 11 *Momordica* species with published chromosome number. The chromosomes observed for all studied species are small, with similar morphology and are metacentric, in general. *M. charantia* is the species with more presented information, with records of chromosome morphology for different populations, specially the Asian ones. Bhaduri and Bose (1947) could not describe the morphological details because of the reduced size of the chromosomes, but registered a pair of chromosome satellites for *M. charantia*. Trivedi and Roy (1972) showed minimum and maximum chromosome length for the same species (1.8–2.38 μm), but did not describe secondary constrictions or satellites. These results are similar to those presented by Zaman and Alam (2009) for three *M. charantia* cultivars, with small chromosomes (0.9–1.5 μm), but no morphological chromosome data. However, the authors indicate a symmetric and stable karyotype for *M. charantia*, even after species domestication.

In a karyomorphological study with seven *Momordica* species, Bharathi et al. (2011) observed 11 pairs of small chromosomes (0.87–1.82 μm) for *M. charantia*, which were grouped according to their morphology. One pair of group A is of relatively medium size, metacentric, with interstitial secondary constriction in the short arm. The group B has four pairs of metacentric medium to small chromosomes, and six pairs of group C have small chromosomes, predominantly submetacentric. Kausar et al. (2015) studied the karyotypes of 34 accessions of *M. charantia* collected in Bangladesh, India, Pakistan and Thailand. In these accessions, there were observed chromosome counts of $2n = 22$, 44 and 66. The authors registered for these accessions chromosome length varying from 0.8 to 2.9 μm, on average. The chromosome length varies between the studied accessions, and a correlation among ploidy level and chromosome length has been observed, the smallest chromosomes registered in the accessions with higher chromosome number ($2n = 66$). The karyotypic formula registered also varies between accessions, but the karyotypes were symmetric, and the chromosomes observed were predominantly metacentric and submetacentric.

Populations of *M. charantia* were studied in Japan. Kido et al. (2016) analyzed cytogenetically three Japanese cultivars and observed the same chromosome number and morphology previously presented for the Asian populations. The chromosome of the three cited populations varies gradually in length, from 1.3 to 3.0 µm in average, all the pairs metacentric. A pair of chromosome satellites was observed, and for the variety *M. charantia* Shiro-Goya polymorphism between the satellites of the chromosome pair was reported.

Ghosh et al. (2018) observed karyotype characteristics similar to the Japanese variety for two *M. charantia* varieties cultivated in India, although the satellite polymorphism was not observed. *M. charantia* var. *charantia* and *M. charantia* var. *muricata* had chromosome length varying from 1.32 to 3.24 µm, and 1.27 to 3.07 µm, respectively. The length variation may be attributed to technique artifact.

Although *M. charantia* is the most studied species in the genus, because of its importance and agronomic interest, other *Momordica* species were also karyomorphologically described. Bharathi et al. (2011) presented data for *M. charantia* and other six species of the genus, with chromosome number, length, volume (Table 5.2), secondary constrictions (SSC) and karyogram (Fig. 5.1).

The largest chromosomes were observed for *M. cymbalaria*, even though still can be considered small chromosomes. For all the species studied, one pair of secondary constriction was observed; in contrast, two pairs were reported for *M. subangulata* subsp. *renigera*. This species presented diploid number $2n = 56$, a tetraploid based on $x = 14$, justifying the double number of nucleoli organization regions (NOR).

The karyotypes presented for *M. charantia* and *M. balsamina* are the most similar morphologically. The karyotypes are very close, even in the karyotypic formula, total chromatin length (TCL) and average chromosome values. Both

species also present monoecism and are included in the group of the basic chromosome number $x = 11$. In this case, de karyomorphological data are added to the numeric counting and morphological description (Bharathi et al. 2011), in order to ensure the systematic positions of the species.

The results presented for the *M. dioica* by Bharathi et al. (2011) corroborate to those described by Barattake and Patil (2009), on 20 populations of this species, collected in India. The authors registered chromosome length average between 0.6 and 1.3 µm for these populations, and the female chromosomes are slightly smaller than those from the male population. This difference acts on the TCL, which was presented as 13.5 µm to the male population and 12.3 µm to the female population. In both populations, a chromosome pair with secondary constriction and prevalence of metacentric chromosomes was observed.

Although there has been observed a variation in chromosome length between male and female individuals for *M. dioica*, Barattake and Patil (2009) did not describe a chromosome sex dimorphism for the species. All the chromosomes are metacentric, with gradual length variation, resulting in a symmetric karyotype. Chattopadhyay and Sharma (1991) observed the same characteristics for two other dioecious species of Cucurbitaceae, *Trichosanthes dioica* and *Coccinia indica*. For these species, it was also reported a variation in TCL for different sex populations, being the TCL higher for the female population, diverging from the data presented for *M. dioica* by Barattake and Patil (2009). Although there are no published reports on sex chromosome dimorphism for *Momordica*, a pair of chromosomes identified as XY was reported for *Coccinia indica* (Cucurbitaceae) male individuals (Guha et al. 2004) similar to the description for *Coccinia grandis* (Shaina and Beevy 2014). These data indicate the presence of sex chromosome in dioecious species, when it occurs, related to the diversifying sex determination in Cucurbitaceae family. Some species, like *M. dioica* and *Trichosanthes dioica*, do not show chromosome dimorphism, and the control of sex determination in the individuals is set by

Table 5.2 Karyotype data of different species of *Momordica* *Source* Bharathi et al. (2011)

Species	No of SSC	Karyotype formula	Chromosome length (μm)			Chromosome volume (μm³)		Form %
			Range	Total	Mean	Total	Mean	
M. balsamina	2	2A + 10B + 10C	0.65–1.98	28.61 ± ← 1.25	1.30	14.73 ± ← 0.45	0.67	38.55
M. charantia	2	2A + 8B + 12C	0.87–1.72	29.04 ± ← 0.78	1.32	14.95 ± ← 0.36	0.68	40.21
M. dioica	2	2A + 16B + 10C	0.85–2.17	38.53 ± ← 0.95	1.38	19.84 ± ← 0.24	0.71	43.88
M. sahyadrica	2	2A + 18B + 8C	0.73–1.83	37.53 ± ← 0.35	1.34	19.32 ± ← 0.53	0.69	44.42
M. cochinchinensis	2	2A + 14B + 12C	1.16–2.03	46.05 ± ← 1.15	1.64	23.71 ± ← 0.55	0.85	42.56
M. subangulata subsp. *renigera*	4	4A + 14B + 38C	0.52–1.26	51.88 ± ← 1.40	0.93	26.71 ± ← 0.42	0.48	32.26
M. cymbalaria	2	2A + 12B + 4C	1.68–3.59	50.00 ± ← 1.45	2.62	25.75 ± ← 0.55	1.43	42.89

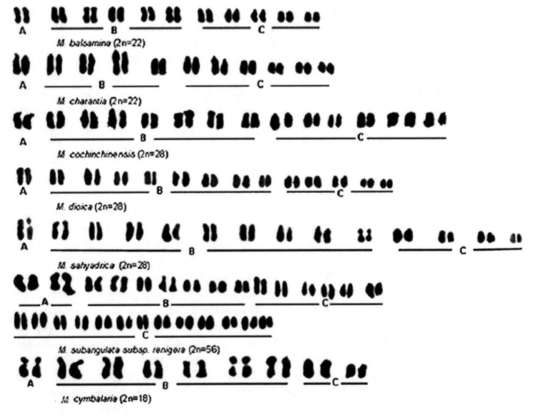

Fig. 5.1 Karyograms of different species of *Momordica*. *Source* Bharathi et al. (2011)

specific genes localized in autosomes (Barrett and Hough 2012).

5.1.3 Heterochromatic Banding and Cytomolecular Analyses

The heterochromatic banding analysis is an important cytogenetic characterization tool, favoring the chromosome pair differentiation in a karyotype, and enables the karyotype distinction for related species (Barros e Silva and Guerra 2010). Banding techniques are used to highlight the heterochromatic sequences, such as the fluorochromes 4′,6-diamidino-2-phenylindole (DAPI) and chromomycin A_3 (CMA) that are widely used to evidence AT and GC rich regions, respectively. These banding techniques are particularly important for closely related species, with similar chromosomes, and symmetric karyotypes, once in these cases it is very difficult to differentiate chromosome pairs. *Capsicum annuum* is one example of the need to use banding techniques, associated with repetitive DNA sequences, in order to differentiate de karyotype, resulting in the distinction of two cultivars (Romero-da Cruz et al. 2017).

Lombello and Pinto-Maglio (2007) analyzed the Brazilian population of *M. charantia* using fluorochrome banding; there were observed four sites CMA+/DAPI−, all of them in the terminal position in the chromosomes. A variation in number of sites CMA+/DAPI− for three cultivated varieties of *M. charantia* was observed by Kido et al. (2016), all cultivars with diploid chromosome number $2n = 22$. Two of these varieties show four sites and the others only one site. Ghosh et al. (2018) identified four distal sites in *M. charantia* var. *charantia* and six in *M. charantia* var. *muricata*. Other six weak signals of CMA+/DAPI − were described in proximal regions for all the studied varieties (Ghosh et al. 2018).

This variation in CMA+ sites observed between cultivated varieties of *M. charantia* may be related to chromosome alterations based on the sequence duplication and posterior dispersion through karyotype, by the action of transposons or retrotransposons, important elements of the karyotypic organization in plants (Heslop-Harrison and Schwarzacher 2011).

These transposable elements can be responsible for the great genetic variability observed by Behera et al. (2008) between the varieties of *M. charantia* var. *charantia* and *M. charantia* var. *muricata* in random amplified polymorphic DNA (RAPD) and microsatellite markers. There was no DAPI positive banding description for *M. charantia* in any of the studied populations. However, for other Cucurbitaceae genus, these heterochromatic regions rich in AT basis can occur. For the cultivated Cucurbitaceae *Benincasa hispida*, *Luffa cylindrica* and *Trichosanthes dioica*, Bhowmick and Jha (2015) observed telomeric DAPI + bands in all analyzed individuals of *T. dioica*, although this was not observed for the other two studied species.

Besides the fluorochrome banding, it can be pointed out that the assays for specific DNA sequences localization using in situ fluorescent hybridization (FISH) were also performed. The sequences to be mapped are those with chromosome marker value, once they are ubiquitous and show variation of number and localization among species' karyotypes. rDNA sequences are largely used in karyotype characterization, especially in species with symmetrical karyotypes or similarly sized chromosome (Chester et al. 2010). The probe of 45S rDNA, complementary to the NOR region, and the probe of 5S rDNA are the most commonly used, as mentioned by Roa and Guerra (2015).

Some cytomolecular studies were carried out with *Momordica*. Lombello and Pinto-Maglio (2007) observed for *M. charantia* four 45S rDNA sites, colocalized with CMA+ sites, and two 5S rDNA sites in proximal position (Fig. 5.2). The colocalization of the CMA+ sites and NOR seems to be a tendency in vegetal karyotype, as described by Roa and Guerra (2012).

Similar results were obtained by Waminal and Kim (2012) in a Korean population of *M. charantia*. The authors identified the 45S rDNA sites in the terminal position of the small arm of chromosome pairs 4 and 11 and the 5S rDNA

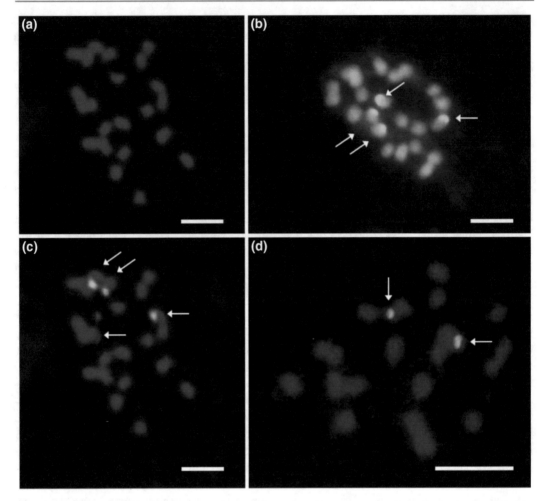

Fig. 5.2 Cells in mitotic metaphase of *M. charantia* stained with: **a** DAPI, **b** CMA, with arrows showing positive bands, **c** hybridized with pTa71 probe for 45S rDNA and counterstained with DAPI and detected with IgG-FITC, arrows showing sites, **d** hybridized with pScT7 probe for 5S rDNA, counterstained with DAPI and detected with IgG-FITC, arrows showing sites. Bar = 2 μm. *Source* Lombello and Pinto-Maglio (2007)

sites in proximal position of the chromosome pair 5 (Fig. 5.3).

The literature presents chromosome mapping studies for sequences of DNA in some Cucurbitaceae species, such as *Benincasa hispida, Citrullus lanatus, Coccinia grandis, Cucumis sativus, Lagenaria siceraria* and *Luffa cylindrica* (Waminal et al. 2011; Waminal and Kim 2012; Sousa et al. 2013; Bhowmick et al. 2016; Li et al.

2016), and for all cited species although there is a variable number of sites, the distribution pattern is the same that observed for *Momordica*, with 45S rDNA sites in the terminal region of the small arm and 5S rDNA in proximal sites. Future studies with species of the *Momordica* genus may identify a numeric pattern of sites, helping the identification of the process of karyotype alteration most important for the genus evolution.

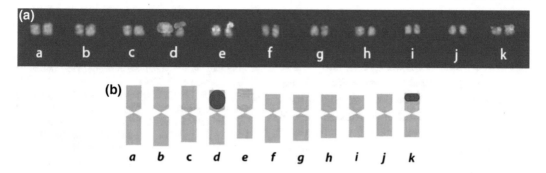

Fig. 5.3 Dual-color FISH mapping and karyotype (**A**) and idiogram (**B**) of *M. charantia*, 5S and 45S rDNA are shown in green and red signals, respectively. *Source* Waminal and Kim (2012)

5.1.4 Meiotic Process and Reproductive Analysis

The analyses of sporogenesis and gametogenesis phases, particularly those using pollinic bags in anthers, bring a great number of information on the studied genome. Other basic information for cytogenetic studies, phylogenetic systematics, reproductive biology and management of economic interest species is the haploid chromosome number counting, the observation of homologous chromosomes pairing and segregation and posterior chromatid distribution in the meiotic pollen mother cell (PMC), the estimation of regular tetrad percentage (meiotic index) and the viable pollen index (Pagliarini 2000; Kirsten 2016; Fernandes et al. 2018).

Some studies carried in *Momordica* species had the initial objective of haploid chromosome counting identification, and also look for pollen viability of cultivated species and hybrids, as well as interspecific genomic compatibility, for genetical enhancement. Trivedi and Roy (1972) analyzed meiosis of *M. charantia*, *M. balsamina* and *M. dioica* and registered regular meiosis for the three species. The monoecious species *M. charantia* ($n = 11$) and *M. balsamina* ($n = 11$) showed similar meiotic process, with the formation of 11 bivalents, and similar chiasm frequency in prophase I. The species *M. dioica* ($n = 14$) showed also only bivalents in metaphases I, however with a lower chiasm frequency than observed for the monoecious species, in other words, lower frequency of genetic recombination. Similar results for *M. balsamina*, with 11 bivalent and regular meiosis, were presented by Janse van Rensburg et al. (1986) in the study of the South African population. Beevy and Kuriachan (1996) studied the meiosis of three *Momordica* species, *M. charantia*, *M. dioica* and *M. denudata*, confirming the previous counts for the first two species mentioned and presenting the unpublished data $n = 14$ for *M. denudata*, with only bivalent formation in metaphases I. Only bivalents were also observed for *M. cochinchinensis* ($n = 14$) and *M. dioica* ($n = 14$) by Bharathi et al. (2010). However, the authors registered for para *M. subangulata* subsp. *renigera* ($2n = 58$) the occurrence of a quadrivalent in diakinesis. In the interspecific reproductive assays, Bharathi et al. (2010) observed for the hybrid *M. dioica* × *M. subangulata* subsp. *renigera* the average formation of 13.84 univalents, 12.76 bivalents and 0.88 trivalents. For the hybrids *M. cochinchinensis* × *M. subangulata* subsp. *renigera*, the authors registered on average 12.96 univalents, 13.08 bivalents and 0.96 trivalents. The observation of the hybrids *M. dioica* × *M. cochinchinensis* resulted in the average formation of 9.12 bivalents in metaphases I, and the other univalents. The formation of various anomalies, such as uni, tri, quadri or multivalent, indicates a failure in homologous chromosome pairing that can result in the formation of micronuclei or unbalanced gametes, reducing pollen fertility of the individuals.

Analyzing the crossing between the varieties of cultivated *M. charantia* (*M. charantia* var.

charantia—MCC) and wild (*M. charantia* var. *muricata*—MCM), Bai and Beevy (2012) did not observe meiotic abnormalities and registered high pollen viability for the hybrid, above 83%. Bano et al. (2019) analyzed the meiotic process in other populations of the same varieties studied by Bai and Beevy (2012) and showed meiotic irregularities in both varieties. For populations of the cultivated variety MCC, the authors registered the diploid number $2n = 22$ for most of the PMC; however, some observed cells were hypoploid (4.32%), with chromosome number $2n = 20$, and hyperploid (1.93%), with $2n = 24$, probably to failure in mitosis that originated microsporocytes. The authors also observed for MCC individuals that showed quadrivalents in metaphases I and had the viability reduced if compared to individuals that showed only bivalents in this phase. The same Indian variety was studied by Ghosh et al. (2018), and no meiotic abnormality was described for MCC or MCM, with the formation of 11 bivalents in both populations, that could be distinguished only for chromosome pair numbers associated with the nucleolus in diakinesis. For MCC, two bivalents were registered associated with the nucleolus, and for MCM three bivalents were observed in the same condition. This positioning, that indicates the number of chromosome pairs with NOR, was confirmed in the study of Ghosh et al. (2018) with the analyses of CMA+ bands concomitant to meiosis observations.

These data support the importance of the meiotic analyses as the scientific basis for reproductive and evolutionary studies, for both wild and cultivated populations, as well as the hybrids, particularly to species of economic, agronomic or medicinal interest, like those of the *Momordica* genus.

References

Agarwal PK, Roy RP (1976) Natural polyploids in Cucurbitaceae I. Cytogenetical studies in triploid *Momordica dioica* Roxb. Caryologia 29(1):7–13

Auquier P, Renard R (1975) Nombres chromosomiques de quelques angiospermes du Rwanda, Burundi et Kivu (Zaïre) I. Bull Jard Bot Belg 45:421–445

Ayyangar KR, Sampathkumar R (1978) On the chiasma specificity in the genus *Momordica*. Proc Indian Sci Congr Assoc 65(111):107

Ayyangar KR (1976) Karyo-taxonomy, karyo-genetics and karyogeography of Cucurbitaceae. In: Nair PKK (ed) Aspects of plant sciences. Today & Tomorrow's Printers & Publishers, New Delhi, India, pp 85–116

Bai NH, Beevy SS (2012) Characterization of intraspecific F1 hybrids of *Momordica charantia* L. based on morphology, cytology and palynology. Cytologia 77(3):301–310

Bano M, Sharma G, Bhagat N (2019) Cytomorphological evaluation of cultivated and wild bitter gourds (*Momordica charantia* L.) of Jammu Province. Natl Acad Sci Lett 42(2):169–173

Barattake RC, Patil CG (2009) Identification of a RAPD marker linked to sex determination in *Momordica dioica* Roxb. Indian J Genet Pl Br 69(3):254–255

Barrett SC, Hough J (2012) Sexual dimorphism in flowering plants. J Exp Bot 64(1):67–82

Barros e Silva AE, Guerra M (2010) The meaning of DAPI bands observed after C-banding and FISH procedures. Biotech Histochem 85(2):115–125

Behera TK, Singh K, Staub JE (2008) Comparative analysis of genetic diversity in Indian bitter gourd (*Momordica charantia* L.) using RAPD and ISSR markers for developing crop improvement strategies. Sci Hort 115(3):209–217

Beevy SS, Kuriachan P (1996) Chromosome numbers of south Indian Cucurbitaceae and a note on the cytological evolution in the family. J Cytol Genet 31:65–71

Bhaduri PN, Bose PC (1947) Cytogenetical investigation in some cucurbits, with special reference to fragmentation of chromosomes as a physical basis of speciation. J Genet 48:237–256

Bharathi LK, Munshi AD, Behera TK, Kattukunnel JJ, Das AB (2010) Cyto-morphological evidence for segmental allopolyploid origin of Teasle gourd (*Momordica subangulata* subsp. *renigera*). Euphytica 176(1):79–85

Bharathi LK, Munshi AD, Chandrashekaran S, Behera TK, Das AB, John KJ (2011) Cytotaxonomical analysis of *Momordica* L (Cucurbitaceae) species of Indian occurrence. J Genet 90(1):21–30

Bhowmick BK, Jha S (2015) Differential heterochromatin distribution, flow cytometric genome size and meiotic behavior of chromosomes in three Cucurbitaceae species. Sci Hort 193:322–329

Bhowmick BK, Yamamoto M, Jha S (2016) Chromosomal localization of 45S rDNA, sex-specific C values, and heterochromatin distribution in *Coccinia grandis* (L.) Voigt. Protoplasma 253(1):201–209

Chattopadhyay D, Sharma AK (1991) Chromosome studies and nuclear DNA in relation to sex difference and plant habit in two species of Cucurbitaceae. Cytologia 56:409–417

Chester M, Leitch AR, Solti PS, Soltis DE (2010) Review of the application of modern cytogenetic methods (FISH/GISH) to the study of reticulation (polyploidy/hybridization). Genes 1(2):166–192

De Storme N, Mason A (2014) Plant speciation through chromosome instability and ploidy change: cellular mechanisms, molecular factors and evolutionary relevance. Curr Plant Biol 1:10–33

Fernandes JB, Seguéla-Arnaud M, Larchevêque C, Lloyd AH, Mercier R (2018) Unleashing meiotic crossovers in hybrid plants. Proc Natl Acad Sci USA 115(10):2431–2436

Ghosh I, Bhowmick BK, Jha S (2018) Cytogenetics of two Indian varieties of *Momordica charantia* L. (bittergourd). Sci Hort 240:333–343

Guha A, Sinha RK, Sinha S (2004) Average packing ratio as a parameter for analyzing the karyotypes of dioecious cucurbits. Caryologia 57(1):117–120

Heslop-Harrison JS, Schwarzacher T (2011) Organization of the plant genome in chromosomes. Plant J 66 (1):18–33

Janse van Rensburg H, Robbertse PJ, Small JGC (1986) Morphology of the anther, microsporogenesis and pollen structure of *Momordica balsamina*. South Afr J Bot 51(2):15–132

Kausar N, Yousaf Z, Younas A, Ahmed HS, Rasheed M, Arif A, Rehman HA (2015) Karyological analysis of bitter gourd (*Momordica charantia* L., Cucurbitaceae) from Southeast Asian countries. Plant Genet Resour Characteriz Utiliz 13(2):180–182

Kirsten L (2016) Meiosis in autopolyploid and allopolyploid *Arabidopsis*. Curr Opin Plant Biol 30:116–122

Kido M, Morikawa A, Saetiew K, Hoshi Y (2016) A cytogenetic study of three Japanese cultivars of *Momordica charantia* L. Cytologia 81(1):7–12

Li K, Wang H, Wang J, Sun J, Li Z, Han Y (2016) Divergence between *C. melo* and African *Cucumis* species identified by chromosome painting and rDNA distribution pattern. Cytogenet Genome Res 150 (2):150–155

Lombello RA, Pinto-Maglio CAF (2007) Cytomolecular studies in *Momordica charantia* L. (Cucurbitaceae), a potential medicinal plant. Cytologia 72(4):415–418

Luceño M, Guerra M (1996) Numerical variations in species exhibiting holocentric chromosomes: a nomenclatural proposal. Caryologia 49(3–4):301–309

Mandáková T, Schranz ME, Sharbel TF, de Jong H, Lysak MA (2015) Karyotype evolution in apomictic *Boechera* and the origin of the aberrant chromosomes. Plant J 82(5):785–793

Mangenot S, Mangenot G (1962) Enquête sur les nombres chromosomiques dans une collection d'espèces tropicales. Bull Soc Bot Fran 109(sup2):411–447

Mehetre SS, Thombre MV (1980) Meiotic studies in *Momordica cymbalaria* Fenzl. Curr Sci 49:289

Mohanty CR, Maharana T, Tripathy P, Senapati N (1994) Interspecific hybridization in *Momordica* species. Myso J Agri Sci 28:151–156

Pagliarini MS (2000) Meiotic behavior of economically important plant species: the relationship between fertility and male sterility. Genet Mol Biol 23 (4):997–1002

Ramsey J, Schemske DW (1998) Pathways, mechanisms, and rates of polyploid formation in flowering plants. Annu Rev Ecol Syst 29(1):467–501

Rice A, Glick L, Abadi S, Einhorn M, Kopelman NM, Salman-Minkov A, Mayzel J, Chay O, Mayrose I (2014) The chromosome counts database (CCDB)—a community resource of plant chromosome numbers. New Phytol https://doi.org/10.1111/nph.13191. Accessed 17 Dec 2018

Richharia RH, Ghosh PN (1953) Meiosis in *Momordica dioica* Roxb. Curr Sci 22:17–18

Riley HP (1960) Chromosomes of some plants from the Kruger National Park. J South Afr Bot 26:37–44

Roa F, Guerra M (2012) Distribution of 45S rDNA sites in chromosomes of plants: structural and evolutionary implications. BMC Evol Biol 12(1):225–237

Roa F, Guerra M (2015) Non-random distribution of 5S rDNA sites and its association with 45S rDNA in plant chromosomes. Cytogenet Gen Res 146(3):243–249

Romero-da Cruz MV, Urdampilleta JD, Forni-Martins ER, Moscone EA (2017) Cytogenetic markers for the characterization of *Capsicum annuum* L. cultivars. Plant Biosyst 151(1):84–91

Roy RP, Thakur V, Trivedi RN (1966) Cytogenetical studies in *Momordica*. J Cytol Genet 1:30–40

Sen R, Datta KB (1975) Sexual dimorphism and polyploidy in *Momordica* L. (Cucurbitaceae). In Proceedings of the 62nd Indian Science Congress. Part III, pp 127

Shaina TJ, Beevy SS (2014) Chromosomal variations in *Coccinia grandis* (L.) Voigt, an actively evolving dioecious cucurbit exhibiting floral polymorphism. Nucleus 57(2):121–127

Shibata K (1962) Estudios citologicos de plantas silvestres y cultivadas. J Agri Sci 8:49–62

Singh AK (1979) Cucurbitaceae and polyploidy. Cytologia 44(4):897–905

Soltis PS, Marchant DB, Van de Peer Y, Soltis DE (2015) Polyploidy and genome evolution in plants. Curr Opin Genet Dev 35:119–125

Sousa A, Fuchs J, Renner SS (2013) Molecular cytogenetics (FISH, GISH) of *Coccinia grandis*: a ca. 3 myr-old species of Cucurbitaceae with the largest Y/autosome divergence in flowering plants. Cytogenet Genome Res 139(2):107–118

The Plant List (2010) Version 1. Published on the Internet; http://www.theplantlist.org/. Accessed 15 Dec 2018

Trivedi RN, Roy RP (1972) Cytological studies in some species of *Momordica*. Genetica 43:282–291

Trivedi RN, Roy RP (1973) Cytogenetics of *Momordica charantia* and its polyploids. Cytologia 38:317–325

Waminal NE, Kim NS, Kim HH (2011) Dual-color FISH karyotype analyses using rDNAs in three Cucurbitaceae species. Genes Genom 33(5):521

Waminal NE, Kim HH (2012) Dual-color FISH karyotype and rDNA distribution analyses on four Cucurbitaceae species. Hort Environ Biotechnol 53(1):49–56

Whitaker TW (1933) Cytological and phylogenetic studies in the Cucurbitaceae. Bot Gaz 94(4):780–790

Zaman MY, Alam SS (2009) Karyotype diversity in three cultivars of *Momordica charantia* L. Cytologia 74:473–478

Sex Determination in Bitter Gourd

6

Hideo Matsumura, Naoya Urasaki, Sudhakar Pandey
and K. K. Gautam

Abstract

Sex expression in the bitter gourd is mainly categorized into monoecious—staminate and pistillate flowers are produced separately in the same plant. Genes for its sex determination mechanism are quite important for genetic improvement through breeding, including the production of hybrid cultivars. In Cucurbitaceae, the mechanism of physiology and genetics of sex determination and expression in cucumber and melon have been extensively studied. For elucidating on sex determination in bitter gourd, a few gynoecious bitter gourd lines, showing only the female flowers, are found and employed as the maternal parent of F_1 hybrid cultivars. Each of these gynoecious lines was predicted to be determined by a single recessive locus, which was genetically mapped. In monoecious bitter gourd, various ratios between female and male flowers per a plant (sex ratio) were observed and frequency of female flowers is an influential trait for the yield of fruits. By genetic mapping approaches, quantitative trait loci for female flower frequency were found. It was already well known that ethylene signaling is a key of sex determination in melon and cucumber, and causal genes for their sex determination were identified. According to bitter gourd draft genome sequences, putative orthologs of these sex determination genes could be identified. In other *Momordica* species, both monoecious and dioecious species were diverged. Genetic and genomic studies of sex determination in bitter gourd will greatly contribute to elucidate the evolution of monoecy and dioecy.

6.1 Introduction

Momordica charantia L. generally known as bitter gourd, bitter melon, or balsam pear and belongs to the family Cucurbitaceae. Bitter gourd is an annual herb which is cultivated worldwide in tropical and subtropical regions of Asia, Africa and South America (Basch et al. 2003; Grover and Yadav 2004). Maximum higher plants (angiosperms) are hermaphrodite (bisexual) species and about ten percent of angiosperms have unisexual flowers. Out of 10% species, half of them are monoecious, where male and female organs

H. Matsumura (✉)
Gene Research Center, Shinshu University, Ueda, Nagano, Japan
e-mail: hideoma@shinshu-u.ac.jp

N. Urasaki
Okinawa Agricultural Research Center, Itoman, Okinawa, Japan
e-mail: uraskiny@pref.okinawa.lg.jp

S. Pandey · K. K. Gautam
ICAR-Indian Institute of Vegetable Research, Varanasi, India
e-mail: sudhakariivr@gmail.com

© Springer Nature Switzerland AG 2020
C. Kole et al. (eds.), *The Bitter Gourd Genome*, Compendium of Plant Genomes,
https://doi.org/10.1007/978-3-030-15062-4_6

coexist in the same individual plant, and the remaining half are dioecious species, where each individual plant has only male or female flowers. Variation in sex forms is observed in cucurbitaceous crops ranges from primitive hermaphrodite to the advanced monoecious sex forms (Robinson and Decker-Walters 1999). The modification of sex from hermaphrodite to the intermediate sex form is mainly due to the evolutionary changes in later generation and dominant mutation effects. These various flowering patterns of sexual expression in bitter gourd are advantageous for exploring the general phenomenon of sex determination. In the early stage of floral bud differentiation, every floral bud has primordia of pistils and stamens, but in later stages, the development of pistil/stamen is arrested in floral buds destined to develop in male/female flower, respectively (Kubicki 1969). Determination of sex in flowering plant is a fundamental developmental process of high economic importance which results in the development of unisexual flowers from an originally bisexual floral meristem (Kater et al. 2001). Sex determination and sex expression are the two important phenomena noticed in cucurbits. Sex expression in cucurbits is a specific gender of individuals in a plant population (Cruden and Lloyd 1995; Neal and Anderson 2005) while sex determination is a developmentally regulated way culminating in plants bearing with different sexual flowers. Sex determination genes control the sex expression, although plant hormones or growth regulators as well as environmental factors (such as day length, humidity and temperature) can alter the expression of the sex-determining genes, leading in many situations to whole sex reversal (Meagher 2007). Production of plant hormone ethylene is more in case of gynoecious cucumber plant than monoecious plant (Rudich et al. 1972).

In bitter gourd, sex expressions are mainly categorized into monoecious (staminate and pistillate flowers are produced separately in the same plant) and gynoecious (only pistillate flowers are produced on a plant), but androecious (only staminate flowers produced), andromonoecy (staminate and perfect flower are produced separately on the same plant), and

hermaphroditic (plant bearing bisexual or perfect flowers) types also exist in nature. Plants owning female and hermaphrodite flowers have also been detected and further used in hybrid development (El-Shawaf and Baker 1981). The mechanism of physiology and genetics of sex determination and expression in cucumber and melon have been extensively studied along with gynoecious sex form which has been commercially exploited worldwide for cucumber improvement programs (Pan et al. 2018).

6.2 Gynoecy in Bitter Gourd

Identical to wild-type monoecious cucumber, in case of bitter gourd also flowers are formed in a preset, developmental sequence along the main stem, with an initial phase of staminate flowers, followed by an alternate phase of pistillate and staminate flowers and end by a pistillate flowering phase (Shifriss 1961). For elucidating sex determination in bitter gourd, mutants or varieties showing altered sex determination were quite important materials, as shown in sex determination studies in melon or cucumber. As genetic mutants, a few gynoecious bitter gourd lines, showing only the female flowers, are found randomly in nature (Zhou et al. 1998; Ram et al. 2002a, b; Behera et al. 2006; Iwamoto and Ishida 2006; Matsumura et al. 2014). These gynoecious lines are quite useful as the maternal parents for F_1 production, like male sterile lines in hybrid seed production of cereals or vegetable crops. The succeeding generations (including F_1) using gynoecious line as one parent showed a very high percentage of female flowers with high yield potential (Ram et al. 2002a; Behera et al. 2006; Iwamoto and Ishida 2006).

In Okinawa, a gynoecious line (OHB61-5) was identified, which was supposed to be spontaneous mutant. According to genetic study of a segregated F_2 population derived from OHB61-5 x OHB95-1A (monoecy), gynoecy was predicted to be determined by a single recessive locus (Matsumura et al. 2014; Table 6.1). Segregation of gynoecy and monoecy was also evaluated in different F_2 populations using OHB61-5 as the

Table 6.1 Segregation of gynoecy in F_2 progeny

	Total analyzed F_2 plants	Monoecious plants	Gynoecious plants	Ratio
OHB61-5 x OHB95-1A	49	37	12	3.08:1
OHB61-5 x OHB61-2	160	123	37	3.32:1
OHB61-5 x OHB1-1	160	119	41	2.90:1

maternal line and OHB61-2 or OHB1-1 as the paternal monoecious line (Table 6.1). In both the population, sexual phenotypes were segregated at 3:1 (monoecy: gynoecy), showing Mendelian inheritance.

Using F_2 population from OHB61-5 x OHB95-1A, a new linkage map was developed based on 584 codominant RAD-seq markers and five single nucleotide polymorphism (SNP) loci were shown to link to gynoecy (Matsumura et al. 2014). The closest SNP marker (*GTFL-1*) was located at 5.46 cM distance to putative gynoecious locus. According to the linkage map, *GTFL-1* was located in the end of the linkage group, and in the draft genome sequence, *GTFL-1* sequence was also positioned in the end region of scaffold 326. By genotyping of 160 F_2 individuals from OHB61-5 x OHB95-1A using *GTFL-1* marker, it was shown that this marker was effective to select gynoecy in the population. However, so far for genotyping bitter gourd resources in Okinawa, this SNP in *GTFL-1* was not observed between OHB61-5 and several monoecious lines (data not shown), indicating additional gynoecy-linked markers or gynoecious gene should be identified for application in marker-assisted breeding.

Other than OHB61-5, gynoecious lines were reported like Gy263B, DBGY-201, or DBGY-202 (Ram et al. 2006; Behera et al. 2006). In these lines, gynoecy was predicted to be determined by a single recessive locus. Gangadhara Rao et al. (2018) mapped gynoecious locus (*gy-1*) using a F_2 population derived from DBGY-201 and a monoecious line. It was located at the middle position of linkage group 12 in their own developed linkage map. Correspondence between two different linkage maps (Urasaki et al. 2017; Gangadhara Rao et al. 2018) was unclear, since the approximate position of each locus in the

linkage group was different, the gynoecious gene in OHB61-5 and DBGY-201 could be independent. On the other hand, Cui et al. (2018) also mapped gynoecy by using a F_2 progeny from gynoecious K44 x Dali-11. Interestingly, two closely linked loci for gynoecy were identified as *gy1.1* and *gy1.2* at the end of linkage group MC01, and *GTFL-1* was also located in the scaffold sequence corresponding to *gy1.1* and *gy1.2* loci. These results demonstrated that causal gene(s) was present in this region. However, gynoecy of OHB61-5 was determined by a single recessive locus, but two loci were concerned in gynoecy of K44 and their LOD scores were significant but not so high (4.43-6.46). Overall, the inheritance of gynoecism has different opinions among researchers, and the majority concluded that gynoecism trait is governed by a monogenic recessive gene (*gy-1*) (Ram et al. 2006) and remaining claiming that gynoecism is partially dominant (Iwamoto and Ishida 2006) or semi-dominant, whereas two pairs of genes were reported by Cui et al. (2018). Further, the study of these loci must be supportive for understanding the mechanism of gynoecious sex expression in bitter gourd.

6.3 Sex Ratio

In monoecious bitter gourd varieties, various ratios between female and male flower per a plant (sex ratio) were observed. Frequency of female flowers is an influential trait for yield of fruits. In F_1 cultivars developed by crossing inbred lines, stable sex ratio was observed (data not shown), indicating that it was genetically determined. Female flower frequency was widely segregated in F_2 progeny of these cultivars, suggesting the presence of quantitative trait loci (QTLs).

Two QTL mapping approaches of sex ratio were reported using F_2 or $F_{2:3}$ population from Z-1-4 and 189-4-1 (Wang and Xiang 2013) or DBGY-201 and Pusa Do Mousami (Gangadhara Rao et al. 2018), resulting in the detection of three or seven QTLs, respectively. Since these results suggested that multiple minor loci affected sex ratio, combination of QTLs must be carefully considered for developing cultivars with appropriate sex ratio.

6.4 Candidate Genes for Sex Determination

Three major loci *F*, *M*, and *A* determined the type of sex in cucumber (Galun 1962; Shifriss 1961; Kubicki 1969). *F* locus impacts on the degree of femaleness (*FF* > *Ff* > *ff*), while the *M* locus governs whether flowers are unisexual (*M-*) or bisexual (*mm*). If a plant is in homozygous recessive condition (*aa* and *ff*), *A* locus increased the male tendency. The interactions between these loci result in different basic sex types in cucumber (Staub et al. 2008). In case of melon, two major loci *A* (andromonoecious) and *G* (gynoecious) determine the sex expression (Kenigsbuch 1990; Li et al. 2012). If plants have the dominant allele at both loci (*A-G-*), it results in monoecious plants, whereas those with a recessive homozygosity at either the *g* or a locus (*A-gg* or *aaG-*) exhibit gynoecy or andromonoecy, respectively. Recessive homozygosity at both loci (*aagg*) results in hermaphroditic flowers.

For sex determination in the *Cucumis* species, it was known that ethylene is a key factor. Although ethylene as a phytohormone was well known to have multiple functions in higher plants, including fruit ripening, senescence, or biotic/abiotic stress response, its role in sex determination was uniquely observed in Cucurbitaceae. Ethylene also played an important role in sex determination in bitter gourd, since treatment by silver nitrate as an inhibitor of ethylene-induced transition of female flower to hermaphrodite flower (Matsumura et al. 2014). Similar effect of silver nitrate treatment was also observed in female flower of dioecious *Momordica dioica* (Hossain et al. 1996), suggesting that ethylene function in sex determination was conserved in *Momordica* species.

The data of molecular characterization of '*A*' and '*G*' loci in melons and '*F*' and '*M*' loci in cucumber directly indicate a significant role of ethylene in sex determination of cucurbits (Pan et al. 2018). In ethylene biosynthesis genes in higher plants, aminocyclopropane-1-carboxylic acid (ACC) synthase genes were identified as sex determination genes in melon and cucumber. *CmAcs11* was responsible for female flower determination in melon (Boualem et al. 2015), and *CmAcs-7*, as an additional ACC synthase gene, encoded by stamen inhibiting '*A*' locus was also shown to regulate unisexual flower development (Boualem et al. 2009). The '*G*' locus encodes a WIP transcription factor gene *CmWIP1* that was confined to pistillate primordia of future male flowers. The expression of *CmWIP1* is antagonistic (leads to bisexuality) with *CmACS7* but does not govern the expression of *CmACS7* gene. Their orthologous genes were also identified and functionally conserved in cucumber.

Furthermore, to ethylene biosynthesis genes, many ethylene preceptor's genes, viz. *CsETR1*, *CsETR2*, and *CsERS*, were also identified which have a significant role in cucurbit sex expression. Gunnaiah et al. (2014) had undertaken an *in silico* differential gene expression analysis between gynoecious and monoecious bitter gourd lines to identify probable candidate genes in ethylene biosynthesis.

Phylogenetic analysis was carried out using putative orthologous proteins for *CmAcs-7* and *CmAcs11* in four Cucurbitaceae plants including bitter gourd (Urasaki et al. 2017). In this analysis, *CmAcs-7* and its homologous proteins were separated from *CmAcs11* and its homologs, predicting the functional differentiation of these two groups of ACC synthases. Furthermore, orthologous proteins for *CmAcs-7* or *CmAcs11* were found in the annotated proteins of seven Cucurbitaceae crops (*C. sativus, Citrullus lanatus, Cucurbita pepo, Cucurbita maxima, Cucurbita moschata, Lagenaria siceraria,* and

M. charantia) by BLAST searching. Proteins, showing the highest identity with *CmAcs-7* or *CmAcs11*, were selected at amino acid level (Table 6.2). Between *CmAcs-7* and each orthologous protein, proteins in cucumber and watermelon showed >90% identities, and those in other species also showed high percentages (88–89%). For *CmAcs11*, the identities of proteins in species other than cucumber showed relatively low score (72–82%). This result suggested that *CmAcs-7* and orthologous proteins were conserved, but sequence divergence in Cucurbitaceae was observed in *CmAcs11* at the similar level as seen in other proteins. Furthermore, two proteins MOMC518_1g (encoding *CmAcs-7*-like protein) and MOMC3_649g (encoding *CmAcs11* like protein) were expressed in female flowers more preferentially than male flower buds of bitter gourd plants/genome (Urasaki et al. 2017) and two copies of *CmAcs-7* orthologous proteins were also present in *Cucurbita* species (data not shown). These differences in gene structure might be related to their functions in sex determination. Therefore, the spatiotemporal expression patterns should be analyzed in flower buds to sort out the function in the sex determination of candidate genes in bitter gourd.

Boualem et al. (2009) firstly showed that the andromonoecy results from a mutation in the active site of the enzyme (*CmACS7*). The active enzyme expression inhibits the development of the male organs without affecting carpel development. The sexual transition from monoecy to andromonoecy is controlled by '*a*' gene in melon and '*M*' gene in cucumber. Isolation of andromonoecy *M* gene was done by using a candidate gene approach with genetic and biochemical analysis in cucumber. Their results confirmed that andromonoecious sex type is also due to a loss of ACS enzymatic activity. They also showed the cosegregation of *CsACS2* and its expression role in carpel primordia of female flowers which is like *CmACS7* in melon, i.e., inhibition of stamina development.

Other than genes for ethylene biosynthesis, causal mutation for gynoecy was identified in *CmWip1* transcription factor of melon. This natural and heritable mutation resulted from the addition of a transposon, which is required for DNA methylation to the *CmWip1* promoter. The expression of *CmWip1* transcription factor directly leads to carpel abortion, results in unisexual male flower development, and indirectly suppresses the expression of the andromonoecious gene, *CmACS-7*, to permit stamen development. Therefore, we can say that interaction of *CmACS-7* and *CmWip1* transcription factors control the overall development of male, female, and hermaphrodite flowers (Martin et al. 2009). Although its homolog was also present in the bitter gourd genome, it expressed in the tissue of gynoecious OHB61-5 and any structural difference between OHB61-5 and other monoecious lines was not found in the homolog of *CmWip1*. Orthologous proteins to *CmWip1* were also found in seven cucurbits as shown in *CmAcs-7* or *CmAcs11* (Table 6.2). Its sequence identities with orthologous proteins were around 70% except for that in cucumber (91%). According to sequence alignment of these proteins (Fig. 6.1), regions encoding Zn-finger domain (276–320 in amino acid sequence of *CmWip1*) were conserved, but other regions were diverged. Still, since biological functions of *CmWip1* orthologs other than cucumber were not demonstrated, further studies are necessary for understanding conservation of *Wip1* function.

In case of bitter gourd transcriptome, sequencing of only flower buds could result in accurate information on gynoecy determining genes. However, a few upregulated candidate genes that control sex expression in melon, cucumber, and Arabidopsis can be further validated for their role in regulating sex expression in bitter gourd.

6.5 Dioecy and Monoecy in *Momordica* Species

In genus *Momordica*, both monoecious and dioecious species are present. Detailed genetic mechanism of sex determination in dioecious *Momordica* species like *M. dioica* is still unknown, but segregation ratio of male: female was 1:1 and male plant was presumed to be

Table 6.2 Amino acid sequence identities of orthologous proteins to CmACS-7, CmACS11, and CmWip1 in Cucurbitaceae plants

Species	Cucumis sativus	Citrullus lanatus	Cucurbita pepo	Cucurbita maxima	Cucurbita moschata	Lagenaria siceraria	Momordica charantia
Accession No.	XP_004135814.2	Cla011230	XP_023516904.1	XP_022987725.1	XP_022960845.1	Lsi06G002550.1	XP_022159242.1
Identity to CmACS-7 (%)	98	94	89	88	88	88	88
Accession No.	XP_004142909.2	Cla022653	XP_023546847.1	XP_022996837.1	XP_022962244.1	Lsi08G015540.1	XP_022131648.1
Identity to CmACS11 (%)	92	80	78	79	78	82	72
Accession No.	XP_004142226.2	Cla008537	XP_023513261.1	XP_022985747.1	XP_022944309.1	Lsi01G019280.1	XP_022148817.1
Identity to CmWip1 (%)	91	73	73	71	73	69	67

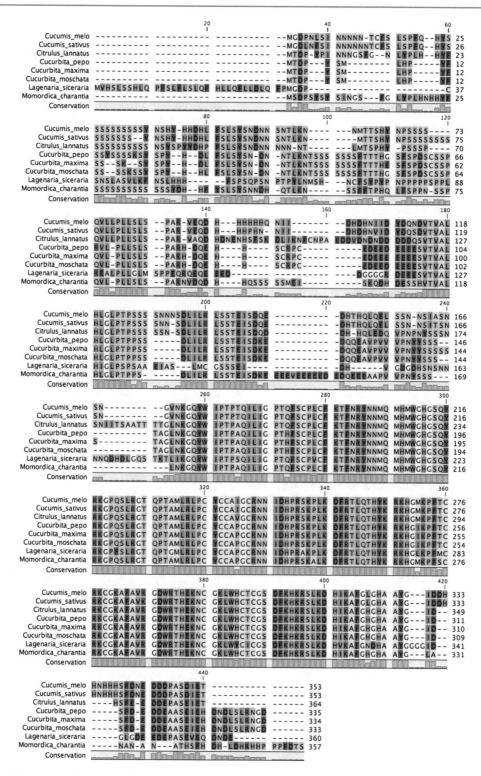

Fig. 6.1 Amino acid sequence alignment of orthologous proteins to CmWip1. Proteins, showing the highest similarity to CmWip1 by BLASTP search against annotated proteins of genome assembly in each species, were selected, and their accession numbers were shown in

Table 6.2. Sequence alignment was carried out by ClustalW. In 'Conservation' below each alignment, conservation rate at each site was indicated by pink-colored bar

heterogametic in *M. dioica* (Hossain et al. 1996). Although dioecious plant species showing such sex determination was likely to have primitive sex chromosome, no heteromorphic chromosomes were seen in both male and female plant of *M. dioica* (Baratakke and Patil 2009).

According to the phylogenetic study of *Momordica* species, seven times of conversion between monoecy and dioecy have occurred during evolution from putative ancestral species (Schaefer and Renner 2010). Boualem et al. (2015) demonstrated the production of artificial dioecious *C. melo* plants by combining alleles of *CmACS11* and *CmWip1*. However, it is difficult to explain natural monoecy–dioecy conversion in *Momordica* species by the combination of these two alleles. Therefore, studies of sex determination in these *Momordica* species greatly contribute to elucidate the evolution of monoecy and dioecy.

References

Baratakke RC, Patil CG (2009) Karyomorphological investigations in dioecious climber *Momordica dioica* Roxb. J Cytol Genet 11:91–96

Basch E, Garbardi S, Ulbricht C (2003) Bitter melon (*Momordica charantia*): a review of efficacy and safety. Amer J Health-Syst Pharm 60(4):356–359

Behera TK, Dey SS, Sirohi PS (2006) DBGy-201 and DBGy-202: two gynoecious lines in bitter gourd (*Momordica charantia* L.) isolated from indigenous source. Indian J Genet 66(1):61–62

Boualem A, Troadec C, Camps C, Lemhemdi A, Morin H, Sari MA, Fraenkel-Zagouri R, Kovalski I, Dogimont C, Perl-Treves R, Bendahmane A (2015) A cucurbit androecy gene reveals how unisexual flowers develop and dioecy emerges. Science 350(6261):688–691

Boualem A, Troadec C, Kovalski I, Sari MA, Perl-Treves R, Bendahmane A (2009) A conserved ethylene biosynthesis enzyme leads to andromonoecy in two *Cucumis* species. PLoS ONE 4(7):e6144

Cruden RW, Lloyd RM (1995) Embryophytes have equivalent sexual phenotypes and breeding systems —why not a common terminology to describe them. Amer J Bot 82:816–825

Cui J, Luo S, Niu Y, Huang R, Wen Q, Su J, Miao N, He W, Dong Z, Cheng J, Hu K (2018) A RAD-based genetic map for anchoring scaffold sequences and identifying QTLs in bitter gourd (*Momordica charantia*). Front Plant Sci 9:477

El-Shawaf IIS, Baker LR (1981) Inheritance of parthenocarpic yield in gynoecious pickling cucumber for once-over mechanical harvest by diallel analysis of six gynoecious lines. J Amer Soc Hort Sci 106(3):359–364

Galun E (1962) Study of the inheritance of sex expression in the cucumber. The interaction of major genes with modifying genetic and non-genetic factors. Genetica 32(1):134–163

Gangadhara Rao P, Behera TK, Gaikwad AB, Munshi AD, Jat GS, Boopalakrishnan G (2018) Mapping and QTL analysis of gynoecy and earliness in bitter gourd (*Momordica charantia* L.) using Genotyping-by-Sequencing (GBS) technology. Front Plant Sci 9:1555

Grover JK, Yadav SP (2004) Pharmacological actions and potential uses of *Momordica charantia*: a review. J Ethnopharmacol 93(1):123–132

Gunnaiah R, Vinod MS, Prasad K, Elangovan M (2014) Identification of candidate genes, governing gynoecy in bitter gourd (*Momordica Charantia* L.) by *in-silico* gene expression analysis. Int J Computer Appl 2:5–9

Hossain MA, Islam M, Ali M (1996) Sexual crossing between two genetically female plants and sex genetics of kakrol (*Momordica dioica* Roxb.). Euphytica 90(1):121–125

Iwamoto E, Ishida T (2006) Development of gynoecious inbred line in balsam pear (*Momordica charantia* L.). Hort Res (Jpn) 5(2):101–104

Kater MM, Franken J, Carney KJ, Colombo L, Angenent GC (2001) Sex determination in the monoecious species cucumber is confined to specific floral whorls. Plant Cell 13(3):481–493

Kenigsbuch DYC (1990) The inheritance of gynoecy in muskmelon. Genome 33(3):317–320

Kubicki B (1969) Investigation of sex determination in cucumber (*Cucumis sativus* L.). Genet Polon 10:69–85

Li Z, Wang S, Tao Q, Pan J, Si L (2012) A putative positive feedback regulation mechanism in *CsACS2* expression suggests a modified model for sex determination in cucumber (*Cucumis sativus* L.). J Exp Bot 63(12):4475–4484

Martin A, Troadec C, Boualem A, Rajab M, Fernandez R, Morin H, Pitrat M, Dogimont C, Bendahmane A (2009) A transposon-induced epigenetic change leads to sex determination in melon. Nature 461(7267):1135–1138

Matsumura H, Miyagi N, Taniai N, Fukushima M, Tarora K (2014) Mapping of the gynoecy in bitter gourd (*Momordica charantia*) using RAD-seq analysis. PLoS ONE 9(1):e87138

Meagher TR (2007) Linking the evolution of gender variation to floral development. Ann Bot 100(2):165–176

Neal PR, Anderson GJ (2005) Are 'mating systems' 'breeding systems' of inconsistent and confusing terminology in plant reproductive biology? or is it the other way around?. Plant Sys Evol 250(3-4):173–185

Pan J, Wang G, Wen H, Du H, Lian H, He H, Pan J, Cai R (2018) Differential gene expression caused by the F and M loci provides insight into ethylene-mediated female flower differentiation in cucumber. Front Plant Sci 9:1091

Ram D, Kumar S, Banerjee MK, Kalloo G (2002a) Occurrence, identification and preliminary characterisation of gynoecism in bitter gourd (*Momordica charantia* L.). Indian J Agri Sci 72(6):348–349

Ram D, Kumar S, Banerjee MK, Singh B, Singh S (2002b) Developing bitter gourd (*Momordica charantia* L.) populations with very high proportion of pistillate flowers. Cucurbit Genet Coop Rep 25:65–66

Ram D, Kumar S, Singh M, Rai M, Kalloo G (2006) Inheritance of gynoecism in bitter gourd (*Momordica charantia* L.). J Hered 97(3):294–295

Robinson RW, Decker-Walters DS (1999) Cucurbits. CAB International, Wallingford, Oxford, UK, p 226

Rudich J, Halevy AH, Kedar N (1972) Ethylene evolution from cucumber plants as related to sex expression. Plant Physiol 49(6):998–999

Schaefer H, Renner SS (2010) A three-genome phylogeny of *Momordica* (Cucurbitaceae) suggests seven returns from dioecy to monoecy and recent long-distance dispersal to Asia. Mol Phylogenet Evol 54(2):553–560

Shifriss O (1961) Sex control in cucumbers. J Hered 52 (1):5–12

Staub JE, Robbins MD, Wehner TC (2008) Cucumber. In: Prohens J, Nuez F, (eds) Handbook of plant breeding; Vegetables I: asteraceae, brassicaceae, chenopodiaceae, and cucurbitaceae. Springer, NY, pp 241–282

Urasaki N, Takagi H, Natsume S, Uemura A, Taniai N, Miyagi N, Fukushima M, Suzuki S, Tarora K, Tamaki M, Sakamoto M, Terauchi R, Matsumura H (2017) Draft genome sequence of bitter gourd (*Momordica charantia*), a vegetable and medicinal plant in tropical and subtropical regions. DNA Res 24 (1):51–58

Wang Z, Xiang C (2013) Genetic mapping of QTLs for horticulture traits in a F_{2-3} population of bitter gourd (*Momordica charantia* L.). Euphytica 193 (2): 235–250

Zhou WB, Luo S, Luo JN (1998) An early maturing and high yielding bitter gourd hybrid Culli No. 1. Plant Breed Abstr 68:1002

Tissue Culture, Genetic Engineering, and Nanotechnology in Bitter Gourd

7

Sevil Saglam Yilmaz and Khalid Mahmood Khawar

Abstract

Bitter gourd (*Momordica charantia* L.) belongs to the genus *Momordica* that includes 45 species. It is cultivated extensively in tropical, subtropical, and rarely under temperate climates. The plant is valued in various disciplines of life and natural sciences. It is extensively used for culinary purposes. Its extracts are important for the treatment of a number of diseases and ailments in traditional and modern medicinal systems because of the abundance of insulin-like peptides, a mixture of steroidal sapogenins and alkaloids. It is rarely used as an ornamental plant. There are very few reports on systematic research on agronomic, breeding, and biotechnological aspects that curtail the improvement of this crop plant. This chapter reviews available information on biotechnology in a bitter gourd that will help understand the current scenario and help in making plans for improvement of bitter gourd.

S. Saglam Yilmaz (✉)
Department of Agricultural Biotechnology, Faculty of Agriculture, Kirsehir Ahi Evran University, Kirsehir, Turkey
e-mail: saglamsevil@gmail.com

K. M. Khawar
Department of Crop Science, Faculty of Agriculture, Ankara University, Ankara, Turkey
e-mail: kmkhawar@gmail.com

7.1 Introduction

Bitter gourd (*Momordica charantia* L.), family Cucurbitaceae, (Heiser 1979) is an annual or perennial plant. The genus *Momordica* contains 45 species. Out of these roots, leaves, tender shoots, and green fruits of only seven species, namely *M. dioica* Roxb. ex Willd., *M. cochinchinensis* (Lour.) Spreng., *M. charantia* L., *M. subangulata*, *M. cymbalaria* Fenslex Naud., *M. balsamina* L., *M. sahyadrica* Kattuk. & V. T. Antony, are used as vegetable or for culinary purpose (Roy and Chakrabarti 2003) as shown in Fig. 7.1. It is widely cultivated in the countries of South Asia and Southeast Asia, Tropical Africa, and America (Kirtikar and Basu 1994). It rarely grows under Mediterranean climatic conditions of the Aegean region under natural conditions (Pers. Observations Prof. Dr. Khalid Mahmood Khawar).

It is believed that the plant originated in Eastern Asia, possibly Eastern India from Odisha (erstwhile Orissa) to Assam passing through Burma to Thailand (Chakravarty 1990) from where it spread to western Asia, tropical Africa, Central and South America (Brazil), through slave trade (Walters and Decker-Walters 1988; Miniraj et al. 1993; Marr et al. 2004).

Bitter gourd is affected by a number of insect pests and diseases but none of them is of significant importance (Robinson and Decker-Walters 1997).

Fig. 7.1 Schematic representation of the fruits of seven *Momordica* species used for culinary purpose namely **a** *M. charantia* L., **b** *M. dioica* Roxb. *ex* Willd., **c** *M. cochinchinensis* (Lour.) Spreng., **d** *M. subangulata* Blume, **e** *M. cymbalaria* Fensl *ex* Naud., **f** *M. balsamina* L., **g** *M. sahyadrica* Kattuk. & V. T. Antony

It is very important to collect and maintain wild germplasm for increasing resistance against biotic and abiotic stresses. Behera et al. (2008a) reported a collection of a gynoecious line (DBGy 201) from indigenous wild material of eastern India. They have reported maintenance of the material by sib-mating. They further reported that the researchers are successful in breed lines with a high percentage and yield of pistillate (female) flowers.

It is not significantly affected by photoperiod (Reyes et al. 1994; Lim 1998) and has tolerance against a range of climates (Lim 1998). The plant can be cultivated in humid, hot tropics, subtropics (Reyes et al. 1994), and under temperate climates (Pers. Observations Prof. Dr. Khalid mahmood Khawar) on soils with pH of 6.0–6.7 with moderate (pH 8.0) tolerance against alkaline soils. Although spring, summer, autumn, or mild winter cultivations are reported, it is generally grown during earlier to mid-summer conditions. The plant grows best at 25–30 °C but is not resistant to frost and cool temperature (Larkcom 1991; Desai and Musmade 1998).

The fruits of the plant are very rich in minerals, amino acids, vitamins, proteins and carbohydrates (Miniraj et al. 1993; Desai and Musmade 1998); however, their nutrient contents vary based on types of environmental conditions under which they grow (Kale et al. 1991; Yuwai et al. 1991). The fruit bitterness comes from cucurbitacins (Decker-Walters 1999) and triterpene glycosides (Okabe et al. 1982) and lacking oxygen function at C-il (Neuwinger 1994). They are most commonly consumed as unripe fruit by making bitter gourd chips, stuffing with minced meat, or slices cooked in different types of curries or in fried forms. Sometimes, the seeds are also used as condiments.

Bitter gourd has been used in folk medicine systems of Latin America, China, South Asia (Decker-Walters 1999), Africa, and Western Asia for curing number of diseases since old times (Naseem et al. 1998; Basch et al. 2003; Parray and Islam 2007) and has hypoglycemic characteristics (Khan and Omoloso 1998). It is also reported to promote potential antispermatogenic and antitumor activities (Xue et al. 1998).

The compounds in bitter gourd also contain abortifacient protein including momorcharin. It has sparkling red seeds that are rich in lycopene (Yen and Hwang 1985).

Its extracts are antiulcerogenic, antihepatotoxic, antimicrobial, antioxidant, and antiviral in their characteristics and could reduce blood sugar level in diabetic patients significantly (Raman and Lau 1996), ulcers, infections (Beloin et al. 2005) dysentery, gout, and rheumatism (Subratty et al. 2005). Bitter gourd protein (MRK29) is also used to inhibit HIV (Jiratchariyakul et al. 2001). Similarly, even the whole plant leaf and fruit extractions are also used to treat infections, measles, human worms, fevers, and hepatitis (Behera et al. 2008b).

The bitter gourd is an ignored crop, and major problems in bitter gourd propagation are nonavailability of standardized agronomic techniques and nonavailability of certified seeds. Generally, the farmers collect seeds from the crop of the previous season and sow them next season.

The plant is generally cultivated by direct seeding and rarely through raising of seedlings transplantation from nursery. There is no seed germination problem under tropical conditions; however, the seeds may face dormancy under subtropical and other climatic conditions. The plant grows as a weed on marginal lands under Turkish/Mediterranean conditions. As the plant is not cultivated, therefore, there is a no record of plant cultivation in the State Statistical Institute. The plants growing under natural environment are consumed by stray animals, and the unripe fruits are collected by amateur collectors for sale in rural markets. Remaining fruits perish on vines, and the seeds are dispersed on the surrounding soils that remain dormant until the next growing season, when they germinate under appropriate conditions of temperature and humidity.

Expression of sex is affected by the environment under which *M. charantia* plantlets grow (Wang et al. 1997). Generally, high gynoecy is observed in short-day conditions and proceed to the sixth-leaf stage (Yonemori and Fujieda 1985). Foliar application of growth regulators can also modify sex expression (Ghosh and Basu 1982).

Improvement of this important neglected medicinal crop is highly desirable for increased yield and the development of resistance against a number of insect pests, diseases, and factors like dryness, drought. The resistance can be increased through the integration of traditional breeding or modern biotechnological approaches that could save time and energy.

There are very little reports on systematic research on agronomic and breeding studies of the plant that further curtails the advancement of plant breeding. This chapter includes information on biotechnological advances and future outlooks including information related to important biotechnological developments about bitter gourd direct and indirect organogenesis through callus culture, genetic transformation, advancements in secondary metabolites production or their effects and nanotechnology under different heads.

7.2 Tissue Culture

7.2.1 Micropropagation and Tissue Culture

Plant tissue culture has a significant importance in bitter gourd biotechnology. This technology is widely used for commercial multiplication and improvement because of very high coefficient of multiplication under laboratory conditions. There are number of reports on tissue culture of many cucurbitaceae plants (Venkateshwarlu 2009; Karim and Ahmed 2010; Ugandhar et al. 2011; Rathod 2015).

First tissue culture with multiple shoot induction of bitter gourd was reported by Agarwal and Kamal (2004) and Yang et al. (2004). Agarwal and Kamal (2004) used MS medium, whereas Yang et al. (2004) induced 90.0% shoots on MS medium having 4.0 mg/l 6-BA +2.0 mg/l kinetin to induce green callus with 66.7% frequency on MS medium with 5 mg/l zeatin and 0.5 mg/l kinetin with proliferation coefficient of 5–6. Similarly, Sikdar et al. (2005) induced multiple shoots on immature cotyledonary nodes of lines BGGB1 and BGGB14 on

MS medium using GA3, IAA, IBA, NAA and KIN, BAP. The shoots grew for 5–6 weeks on medium containing 2 mg/l BAP+0.1 mg/l IAA +2 mg/L GA3 to induce adventitious shoot proliferation on immature cotyledonary nodes with 80–84 shoots, after 6 weeks of culture.

Al Munsur MAZ et al. (2007) noticed that root tips were better in the regeneration of callus compared to leaf explants on MS medium with 2.0 mg/l BAP and 0.3 mg/l NAA. They noted >65% shoot regeneration on leaf segments and root tips on medium containing 2.0 or 2.5 mg/l BAP + 0.2 mg/l IAA each. Al Munsur MAZ et al. (2009) used root and nodes explants of bitter gourd cultured on MS medium to induce callus on 1.0 mg/l 2,4-D and 1.0 mg/l BAP with 75.00% shoot regeneration. No regeneration of shoots was noted on root explants in any combinations of plant growth regulators. However, Thiruvengadam et al. (2012a, b) induced shoot regeneration on internodal explants induced calli of bitter gourd on MS medium with 4.0 μM TDZ, 1.5 μM 2,4-D+0.07 mM L-glutamine induced 96.5% (48 shoots per explant). Verma et al. (2014) obtained maximum shoot length on ABG-6 medium (½ MS medium with 0.5 mg/ml BAP after third subculture). Saglam (2017) obtained callus-induced somatic embryogenesis developed on stem explants of accession Silifke genus. The researcher obtained one plantlet on callus induced on the stem explants.

7.2.2 Rooting

Agarwal and Kamal (2004), and Thiruvengadam et al. (2012b) reported rooting of in vitro developed shoots on IBA. However, Huda and Sikdar (2006) have reported rooting in vitro cultured shoots on 0.1 mg/l NAA in MS medium. Al Munsur MAZ et al. (2007) noted no rooting and Al Munsur MAZ et al. (2007) noted the maximum counts (6.75) of roots and with average root length of 2.45 cm on 3.0 mg/L BAP + 0.1 mg/L NAA using root tips, whereas Verma et al. (2014) induced roots on RBG7 medium (½ MS medium with 1.0 mg/l IBA). Yang et al. (2004) used the 1/2×MS + ZT 0.02 mg/l Zeatin or

1/2×MS medium for in vitro rooting of shoots inducing 6–7 new roots in three weeks.

7.2.3 Acclimatization

Agarwal and Kamal (2004) and Verma et al. (2014) reported 40% and full acclimatization and 50% survival of in vitro regenerated plantlets to the external conditions in the fields in the same order, whereas Huda and Sikdar (2006) and Saglam (2017) noted full acclimatization of plants to the external conditions. Yang et al. (2004) noted a survival rate of 70% using field tests. All of the researchers noted no change of morphological characteristics of in vitro regenerated plantlets during field testings.

Paul et al. (2009) induced somatic embryogenesis on leaves of *M. charantia* used as explants after 21 days using MS medium. The researchers observed changes in explants under scanning electron microscopy and found that putrescine was more inductive to multiple cell division and shoot regeneration when compared to spermidine (Spd) and spermine (Spm). They also noted that the addition of polyamines (PAs) to the embryogenic medium significantly improved fresh weight, and somatic embryos count on developing calli after 21 d of culture.

7.2.4 Secondary Metabolite Production

Safdar and Alveena (2013) analyzed secondary metabolites on calli of *M. charantia* cv. Jaunpuri. Different explants from in vitro grown seedlings were utilized and were cultured on MS medium + different PGRs to induce callus. They found that MS medium containing BAP, IAA, 2,4-D BAP, and 2,4-D in almost all combinations was appropriate for the induction of calli. These calli were subjected to GC-MS. The results confirmed that these plant tissues contained significantly more secondary metabolites compared to calli of related explants. The results showed positive effects of secondary metabolites detected on callus cultures of cotyledon explants.

Alpha-eleostearic acid that is known as a momordica specific fatty acid was noted on both internodes of field grown plants and in callus cultures.

Chen et al. (2015) found five cucurbitacins, kuguacins II–VI along with five known analogues to elucidate *M. charantia* with given structure of (i) 5β,19-epoxycucurbit-23-en-7-on-3β,25diol, (ii) 5β,19-epoxycucurbit-7,23-dion-3β,25diol, (iii) 5β,19-epoxycucurbit-6-en-19,23-dion3β,25-diol, (iv) 5β,19-epoxy-23,24,25,26,27-pentan-orcucurbit-6-en-7,19-dion-3β,22-diol, and (v) cucurbit-5-en-7,23-dion-3β,19,25-triol after single-crystal X-ray diffraction and extensive spectroscopic analyses.

Fonseka et al. (2007) quantified hypoglycemic activity on blood sugar reducing effect using a number of bitter gourd cultivars and found that significant glucose reduction percentage in the range of 13.8–35.2%. They noted that glucosidase mediated glucose reduction had a range of 11.9–63.3%. They noted significantly higher anti-glucosidase and anti-amylase activities in green colored fruits compared to white-colored fruits.

7.3 Genetic Transformation

Agrobacterium-mediated β-glucuronidase expression was detected on shoots regenerated on immature cotyledon nodes (Sikdar et al. 2005; Thiruvengadam et al. 2010). The researchers confirmed genetic transformation through histochemical GUS test. Thiruvengadam et al. (2012a) used pre-cultured leaf explants and treated them with A. tumefaciens strain LBA4404 harboring a binary vector pBAL2 carrying gus gene (reporter gene β-glucuronidase) and neomycin phosphotransferase-nptII (marker gene). The putative expected transgenic plants were transferred and acclimatized in greenhouse ensued by the collection of mature fruits. The putative/expected transgenics were validated by histochemical GUS test, classic PCR, and Southern blot test with transformation percentage of

7%. Muralikrishna et al. (2018) used biolistic particle bombardment system for delivery of DNA-coated gold particles (0.6 µm) in transformation. The bacteria harbored binary vector pBI121 along with β-glucuronidase *gus* gene (GUS) and *npt*II gene, respectively. They also confirmed transformants using GUS histochemical assay and by PCR analysis.

7.4 Nanotechnology

Nanotechnology is a new discipline that involves the use of atoms, electrons, protons, and neutrons in number of ways to understand materials to solve problems in geology, life, surface, and marine sciences.

Generally, nanoparticles are made up of copper, gold, iron, palladium, quantum dots (CdS, ZnS), silver, and zinc. A number of experiments report potential effects of nanoparticles on plants. The study reports effects of fullerol on agroeconomic traits in *M. charantia* that resulted in positive increase of fruit yield along with phytomedicinal contents through nanoparticles' interventions (Kole et al. 2013).

Recent research on nanoparticles in a number of crops has evidenced for enhanced germination and seedling growth, physiological activities including photosynthetic activity and nitrogen metabolism, mRNA expression and protein level and also positive changes in gene expression indicating their potential use in crop improvement (Kole et al. 2013).

The green technology of nanoparticle synthesis provides a significant advancement when it is compared to other types of methods (chemical and physical) to commercially synthesize nanoparticles. Dhivya and Rajasimman (2015) and Ekezie et al. (2016) prepared silver and copper nanoparticles (in the same order) prepared by green synthesis that showed significant positive effects on antidiabetic activities and carbohydrate digestibility. Therefore, it was recommended that these nanoparticles should be used as effective agents for the treatment of patients with diabetes.

7.5 Conclusion and Future Prospects

Bitter gourd is a multipurpose plant used as vegetable. It has extensive use in ethnobotany as folk medicine and bitter flavoring agent. Very few studies have been reported on micropropagation, bitter gourd genetic transformation, nanotechnology, and secondary metabolite production. The results suggest that there is a need to focus and make serious inputs for improvement in all fields of bitter gourd biotechnology.

References

Agarwal M, Kamal R (2004) Studies on steroid production using in vitro cultures of Momordica charantia. J Med Arom Plant Sci 26:318–323

Al Munsur MAZ, Haque MS, Nasiruddin KM, Hasan MJ (2007) Regeneration of bitter gourd (Momordica charantia L.) from leaf segments and root tips. Prog Agri 18(2):1–9

Al Munsur MAZ, Haque MS, Nasiruddin KM, Hossain MS (2009) In vitro propagation of bitter gourd (Momordica charantia L.) from nodal and root segments. Plant Tissue Cult Biotechnol 19(1):45–52

Basch E, Gabardi S, Ulbricht C (2003) Bitter melon (Momordica charantia): a review of efficacy and safety. Amer J Health Syst Pharm 60:356–359

Behera TK, Singh AK, Staub JE (2008b) Comparative analysis of genetic diversity in Indian bitter gourd (Momordica charantia L.) using RAPD and ISSR markers for developing crop improvement strategies. Sci Hort 115(3):209–217

Behera TK, Gaikward AB, Singh AK, Staub JE (2008a) Relative efficiency of DNA markers (RAPD, ISSR and AFLP) in detecting genetic diversity of bitter gourd (Momordica charantia L.). J Sci Food Agri 88 (4):733–737

Beloin N, Gbeassor M, Akpagana K, Hudson J, Soussa KD, Koumaglo K, Arnason JT (2005) Ethnomedicinal uses of Momordica charantia (Cucurbitaceae) in Togo and relation to its phytochemistry and biological activity. J Ethnol Pharmacol 96:49–55

Chakravarty HL (1990) Cucurbits of India and their role in the development of vegetable crops. In: Bates DM, Robinson RW, Jeffrey C (eds) Biology and utilization of Cucurbitaceae. Cornell University Press, Ithaca, NY, pp 325–334

Chen JC, Bik-San Lau C, Chan JYW, Fung KP, Leung PC, Liu JQ, Zhou L, Xie MJ, Qiu MH (2015) The antigluconeogenic activity of cucurbitacins

from Momordica charantia. Planta Med 81(04):327–332

Decker-Walters DS (1999) Cucurbits, Sanskrit, and the Indo. Aryas Econ Flot 53:98–112

Desai UT, Musmade AM (1998) Pumpkins, squashes and gourds. In: Salunkhe DK, Kadam SS (eds) Handbook of vegetable science and technology: production, composition, storage and processing. Marcel Dekker Publishers, New York, pp 273–298

Dhivya G, Rajasimman M (2015) Synthesis of silver nanoparticles using Momordica charantia and its applications. J Chem Pharm Res 7:107–113

Ekezie FGC, Jessie Suneetha W, Uma Maheswari K, Prasad TNVKV, Anila KB (2016) Momordica charantia extracts in selected media: screening of phytochemical content and in vitro evaluation of anti-diabetic properties. Indian J Nutr Diet 53(2):164

Fonseka HH, Chandrasekara A, Fonseka RM, Wickramasinghe P, Kumara PDRSP, Wickramarachchi WNC (2007) Determination of anti-amylase and anti-glucosidase activity of different genotypes of bitter gourd (Momordica charantia L.) and thumba karavila (Momordica dioica L.). Acta Hort 752:131–136

Ghosh S, Basu PS (1982) Effect of some growth regulators on sex expression of Momordica charantia. Sci Hort 17:107–112

Heiser CB (1979) The Gourd Book. University of Oklahoma Press, Norman, OK

Huda AKMN, Sikdar B (2006) In vitro plant production through apical meristem culture of bitter gourd (Momordica charantia L.). Plant Tissue Cult Biotechnol 16(1):31–36

Jiratchariyakul W, Wiwat C, Vongsakul M, Somanabandhu A, Leelamanit W, Fujii I, Suwannaroj N, Ebizuka Y, Weena J, Chanpen W, Molvibha V, Somanabandhu A, Leelamanit W, Suwannarol N (2001) HIV inhibitor from Thai bitter gourd. Planta Med 67(4):350–353

Kale AA, Cadakh SR, Adsule RN (1991) Physico-chemical characteristics of improved varieties of bittergourd (Momordica charantia L.). Maharashtra J Hort 5:56–59

Karim MA, Ahmed SU (2010) Somatic embryogenesis and micro propagation in teasle gourd. Int J Environ Sci Dev 1(1):10–14

Khan MR, Omoloso AD (1998) Momordica charantia and Allium sativum: broad spectrum antibacterial activity. Kor J Pharmacog 29:155–158

Kirtikar KR, Basu BD (1994) Momordica charantia Linn. In: Singh B, Signh MP (eds) Indian medicinal plants, vol. II, Dehra Dun. Lalit Mohan Basu, Allahabad, Jayyed Press, New Delhi, India, pp 1130–1132

Kole C, Kole P, Randunu KM, Choudhary P, Podila R, Ke PC, Rao AM, Marcus RK (2013) Nanobiotechnology can boost crop production and quality: first evidence from increased plant biomass, fruit yield and phytomedicine content in bitter melon (Momordica charantia). BMC Biotechnol 13(1):37–47

Larkcom J (1991) Oriental vegetables: the complete guide for garden and kitchen. John Murray, London

Lim TK (1998) Loofahs, gourds, melons and snakebeans. In: Hyde KW (ed) The new rural industries. Rural Industries Research and Development Corporation, Canberra, AU, pp 212–218

Marr KL, Xia YM, Bhattarai NK (2004) Allozyme, morphological and nutritional analysis bearing on the domestication of Momordica charantia L. (Cucurbitaceae). Leon Bot 58:435–455

Miniraj N, Prasanna KP, Peter KV (1993) Bitter gourd Momordica spp. In: Kalloo C, Bergh BO (eds) Genetic improvement of vegetable plants. Pergamon Press, Oxford, UK, pp 239–246

Muralikrishna N, Ellendula R, Kota S, Kalva B, Velivela Y, Abbagani S (2018) Efficient genetic transformation of Momordica charantia L. by microprojectile bombardment. 3 Biotechnol 8(1):2

Naseem MZ, Patil SR, Patil SR, Ravindra, Patil SB (1998) Antispennatogenic and androgenic activities of Momordica charantia (Karela) in albino rats. J Ethnopharmacol 61(1):9–16

Neuwinger HD (1994) African ethnobotany, poisons and drags. Chapman and Hall, London

Okabe H, Miyahara Y, Yamauchi T (1982) Studies on the constituents of Momordica charantia L. III. Chem Pharm But 30:3977–3986

Parray A, Islam A (2007) Bitter gourd (Momordica charantia): a natural gift in support of the research in medicine and biotechnology. J Biotechnol 7(1):1–13

Paul A, Mitter K, Sen RS (2009) Effect of polyamines on in vitro somatic embryogenesis in Momordica charantia L. Plant Cell Tissue Org Cult 97:303–331

Raman A, Lau C (1996) Anti-diabetic properties and phytochemistry of Momordica charantia L. (Cucurbitaceae). Phytomedicine 2:349–362

Rathod V (2015) Plant regeneration in Momordica dioica (Roxb) by root explant. J Pharmacy Biol Sci 10(2):80–83

Reyes MEC, Gildemacher BH, Jansen GJ (1994) Momordica L. In: Siemonsma JS, Piluek K (eds) Plant resources of South-East Asia: vegetables. Pudoc Scientific Publishers, Wageningon, Netherlands, pp 206–210

Robinson RW, Decker-Walters DS (1997) Cucurbits. CAB International, Wallingford, Oxford, UK

Roy SK, Chakrabarti AK (2003) Vegetables of tropical climates/commercial and dietary importance. Encyclopedia of Food Sciences and Nutrition (Second Edition). 5956–5961

Safdar A, Alveena T (2013) Analysis of secondary metabolites in callus cultures of Momordica charantia cv Jaunpuri. Biologia 59(1):23–32

Saglam S (2017) In vitro propagation of bitter gourd (Momordica charantia L.). Sci Bull Sr F Biotechnol 21:46–50

Sikdar B, Shafiullah M, Chowdhury AR, Sharmin N, Nahar S, Joarder OI (2005) Agrobacterium-mediated GUS expression in bitter gourd (M. charantia L.). Biotechnology 4:149–152

Subratty AH, Gurib-Fakim A, Mabomoodally E (2005) Bitter melon: an exotic vegetable with medicinal values. Neu Food Sci 35:143–147

Thiruvengadam M, Praveen N, Chung IM (2012a) An efficient Agrobacterium tumefaciens-mediated genetic transformation of bitter melon (Momordica charantia L.). Austral J Crop Sci 6(6):1094–1100

Thiruvengadam M, Praveen N, Chung III-Min (2012b) In vitro regeneration from internodal explants of bitter melon (Momordica charantia L.) via indirect organogenesis. Afr J Biotechnol 11(32):8218–8224

Thiruvengadam M, Rekha KT, Jayabalan N, Yang CH, Chung IM (2010) High frequency shoot regeneration from leaf explants through organogenesis of bitter melon (Momordica charantia L.). Plant Biotechnol Rep 4:321–328

Ugandhar T, Venkateshwarrlu M, Begum G, Srilatha T, Jaganmohanreddy K (2011) In vitro plant regeneration of cucumber (Cucumis sativum (L.) from cotyledon and hypocotyl explants. Sci Res Rep 1(3):164–169

Venkateshwarlu M (2009) Direct multiple shoots proliferation of muskmelon (Cucumis melo L.) from shoot tip explants. Intl J Pharma Biosci 2(3):645–652

Verma AK, Kumar M, Tarafdar S, Singh R, Thakur S (2014) Development of protocol for micro propagation of gynoecious bitter gourd (Momordica charantia L). Int J Plant Anim and Environ Sci 4(4):275–280

Walters TW, Decker-Walters DS (1988) Balsampear (Momordica charantia, Cucurbitaceae). Econ Bot 42:286–286

Wang Q, Zang GW, Jiang YT (1997) Effects of temperature and photoperiod on sex expression of Momordica charantia. China Vegetables 1:1–4

Xue Y, Song S, Chen H, Xue Y, Song SH, Chen H, Peron JY (1998) Possible anti-tumor promoting properties of bitter gourd and some Chinese vegetables. In: Third international symposium on diversification of vegetable crops, Belling, China (Acta Hort 467:55–64)

Yang M, Zhao M, Zeng Y, Lan L, Chen F (2004) Establishment of in vitro regeneration system of bitter melon (Momordica charantia L.). High Technol Lett 10(1):44–48

Yen GC, Hwang LS (1985) Lycopene from the seeds of ripe bitter melon (Momordica charantia) as a potential red food colorant. II. Storage stability, preparation of powdered lycopene and food application. J Chin Agri Chem Soc 23:151–161

Yonemori S, Fujieda K (1985) Sex expression in Momordica charantia L. Sci Bull Coll Agric Univ Rynkyus, Okinawa 32:183–187

Yuwai KR, Rao KS, Kaluwin J, Jones GP, Rivetts DE (1991) Chemical composition of Momordica charantia L. fruits. J Agri Food Chem 39:1782–1783

Classical Genetics and Traditional Breeding

8

Tusar Kanti Behera, Gograj Singh Jat and Mamta Pathak

Abstract

Much progress has been attempted in classical genetics and traditional breeding of bitter gourd which are mainly related to qualitative traits but significant advancement in several quantitative traits is difficult to achieve. The main purpose of this chapter is to highlight some of the key concepts that lay the genetical foundations for bitter gourd breeding. The study on classical genetics of bitter gourd has added an advantage to the breeders in the development of new varieties and F_1 hybrids for earliness and higher productivity due to the involvement of gynoecious lines as one of the parents in breeding of these varieties. The selection based on morphological traits for high and stable yield requires the evaluation of germplasm in multiple environments over several seasons; which is very expensive and time-consuming process. Molecular markers technology have great potential to overcome many of the obstacles presented by traditional breeding techniques, but it is imperative that the development and utilization of these markers works in conjunction with traditional breeders who have neces-sary skill to evaluate the germplasm lines of economic value. Marker-assisted selection (MAS) certainly accelerates the breeding process and is a powerful tool for selecting for desirable traits. The construction of a genetic map is a common approach to detect quantitative trait loci (QTLs) for genetic improvement of bitter gourd.

8.1 Introduction

Bitter gourd, balsam pear, bitter melon, or bitter cucumber (*Momordica charantia* L.) is an important dicot vine species of the Cucurbitaceae family. It issued as a vegetable crop and also as a medicinal plant in the tropical and subtropical regions of the world. Bitter gourd is believed to be originated in the Indo-Malayan region and has acclimatized widely in the Old and New Worlds (Bates et al. 1995). It is extensively cultivated in India, China, Malaysia, Africa, and South America (Miniraj et al. 1993). The fruits are widely consumed as a vegetable at immature stage which possess medicinal properties such as antidiabetic (Robinson and Decker-Walters 1997; Chen et al. 2003), hypoglycemic (Jayasooriya et al. 2000), anticarcinogenic and hypercholesterolemic (Ganguly et al. 2000; Ahmed et al. 2001), and anti-HIV (Lee et al. 1995) activities. Bitter gourd fruits also contain

T. K. Behera (✉) · G. S. Jat
Division of Vegetable Science, ICAR-Indian Agricultural Research Institute, New Delhi, India
e-mail: tusar@rediffmail.com

M. Pathak
Department of Vegetable Science, Punjab Agricultural University, Ludhiana, India

© Springer Nature Switzerland AG 2020
C. Kole et al. (eds.), *The Bitter Gourd Genome*, Compendium of Plant Genomes,
https://doi.org/10.1007/978-3-030-15062-4_8

charantin (Yeh et al. 2003), momorcharin (Leung et al. 1997), and momordicosides A and B (Okabe et al. 1980). The first one is antidiabetic and the later two are anticancer. The bitterness of bitter gourd fruits is due to the cucurbitacin-like alkaloid momordicine and triterpene glycosides (momordicoside K and L) (Behera et al. 2010). Cucurbitacins have also been reported to possess anticancer activities. Bitter gourd fruits are rich in vitamin C and phenolic compounds with high antioxidant activity (Myojin et al. 2008; Nicoli et al. 2008; Behera et al. 2010; Krishnaiah et al. 2011). Indian bitter gourd has wide phenotypic variation with respect to growth habit, maturity, fruit shape, size, color, surface texture (Robinson and Decker-Walters 1997), and sex expression (Behera et al. 2006). The fruits of bitter gourd are characterized by their warty-skinned fruits. Bitter gourd is among the most popular cucurbits and has very wide commercial distribution in India. High yield and uniform fruit shape, size, and excellent quality are prerequisites for the release of bitter gourd varieties and F$_1$ hybrids. Similar to other cucurbitaceous crops, bitter gourd is a monoecious species. However, some gynoecious lines have been identified in India (Behera et al. 2009), providing useful genetic resources in genetic improvement programs for the production of F$_1$ hybrids. The utilization of gynoecious lines is economical and easier for hybrid seed production (Behera et al. 2009) because it reduces the cost of male flower pinching and hand pollination (Behera et al. 2009). Conventional phenotypic selection for high and stable yield requires the evaluation of yield in multiple environments over several seasons, which is very expensive and time consuming. The scarcity of polymorphic molecular markers in the public database has hindered genetic mapping and the application of molecular breeding in bitter gourd. The molecular basis of agronomically important traits remains unexplored to date and decisive linkage map has not been reported in bitter gourd (Yuan et al. 2002). In contrast, marker-assisted selection (MAS) certainly accelerates the breeding process and is a powerful tool for selecting traits such as

gynoecism for earliness and high yield. In contrast to popular cucurbits, it lacks a high-density genetic linkage map as required for genomic depiction and precise breeding (Kole et al. 2012). Recently, the economic traits like fruit color, fruit luster, fruit surface structure, stigma color, and seed color were mapped and quantitative trait loci (QTLs) controlling several polygenic fruit traits including length, diameter, weight, number, and yield were detected which will be useful for its genetic enhancement (Kole et al. 2012).

8.2 Classical Genetics in Bitter Gourd

8.2.1 Genetics of Different Quantitative and Qualitative Characters

Classical genetics have enriched our understanding of the bitter gourd crop and facilitated the breeders for the development of improved varieties and hybrids. Breeders in the past have been able to make improvement without understanding of the genetic control of these traits, but improvements under these conditions are normally slow. Understanding how deferent genes affect a variety of traits allow breeders to devote the proper resources needed to improve a particular trait. For example, if a breeder is selecting for a trait controlled by a single gene, the population size will likely be much smaller than if the trait is controlled by multiple genes with a large environmental influence. The application of Mendelian genetics using classical techniques has facilitated the discovery of a number of genes and their inheritance in bitter gourd. In bitter gourd, yield may be increased by altering plant architecture to produce gynoecious, early flowering (node and days to first pistillate flower appearance), and cultivars with better sex ratio. In bitter gourd breeding, the number of fruits per plant, fruit weight, and fruit size is the direct yield components.

8.2.1.1 Seed and Fruit Characters

The character like light brown seed (*lbs*) is recessive to dark brown (Srivastava and Nath 1972; Ram et al. 2006; Kole et al. 2012) and large seed (*ls*) is recessive to small seed size (Srivastava and Nath 1972). The monogenic inheritance was reported for fruit color, fruit luster (*FRLsr*), fruit surface structure (*FrSr*), and stigma color (*StCol*) but digenic mode of inheritance was reported for seed color (Kole et al. 2012). Suribabu et al. (1986), Vahab (1989), and Esquinas-Alcazar and Gulick (1983) also reported white epicarp to be recessive to green. Vahab (1989) observed spiny (triangular tubercles) fruit to be dominant over smooth. Since immature bitter gourd fruits are sliced during the preparation of various Asian meals; therefore, internal fruit quality and uniform green peel color are desirable. In addition to appropriate internal color, fruit must be firm, without excessive seed development, and free of internal defects, such as decay and splitting. Fruit color also governs its marketability, although color preference differs among regions. For example, green-fruited types are in demand in southern China, while white-fruited types are preferred in central China. Similarly, dark green to glossy green fruits are favored in northern India, whereas white fruits are preferred in southern India. Liu et al. (2005) reported high heritability of fruit color (green vs. white) controlled by two genes where green is dominant to white (Miniraj et al. 1993; Hu et al. 2002; Liou et al. 2002). Srivastava and Nath (1972), Hu et al. (2002), and Dalamu et al. (2012) studied the inheritance of fruit color in bitter gourd and suggested that the green color is dominant over the white color. The fruit color trait was also reported to be controlled by polygenes (Huang and Hsieh 2017). Srivastava and Nath (1972) found immature fruit color in bitter gourd was controlled by one nuclear gene with no cytoplasmic factor involved. These results implied the inheritance of fruit color and chlorophyll concentrations were controlled by more than three genes, that is, the four color-related traits were quantitative in nature (Dalamu et al. 2012; Huang and Hsieh 2017). The light green colors were probably affected by incomplete dominance or modifier genes (Hu et al. 2002). The continuous variation in fruit length indicated its quantitative inheritance and more than four genes were reported to be involved in controlling this trait (Kumari et al. 2015). The short fruit length is partially dominant over long fruit length, and tubercles and curviness of fruits in bitter gourd, governed by a single pair of nuclear gene and tubercles were dominant over non tubercles (Kumari et al. 2015). The inheritance of ridgeness indicated that the fruit surface (discontinuous ridge) of bitter gourd was governed by a single dominant gene (Dalamu et al. 2012). Fruit yield and its traits are polygenic and exhibit continuous variation (Kole et al. 2012).

8.2.1.2 Sex Expression

The sex expression in bitter gourd has played an important role in seed production as well as development of new fruit types. The flowering traits like days to first pistillate flower appearance, node at first pistillate flower appearance, and male: female ($\male:\female$) flower ratio (sex ratio) are directly related to earliness and fruit yield. Iwamoto and Ishida (2006) reported that gynoecious sex expression was partially dominant in bitter gourd. The gynoecism is governed by a single recessive gene *gy-1* (Ram et al. 2006; Behera et al. 2009; Mishra et al. 2015).Their observations, however, were made using Japanese germplasm (i.e., line LCJ 980120; predominantly female). Both these studies suggested that such gynoecious or predominantly female lines hold promise for the development of gynoecious F_1 hybrids. The gynoecious sex in bitter gourd is governed by two pairs of genes (Cui et al. 2018). The sex is controlled by a single factor, with heterozygous males and homozygous recessive females in *M. dioica* (Hossain et al. 1996) and *M. cochinchinensis* (Sanwal et al. 2011).

8.2.1.3 Yield Characters

Singh and Ram (2005) determined that complementary epistasis and dominance × dominance interactions were important genetic determinates of yield. Given these facts, Devadas and Ramadas (1994) recommended that selection and hybridization (i.e., reciprocal recurrent selection) would be an appropriate breeding strategy for improvement of fruit triterpinoid content. The genetic analysis of a large-fruited (*M. charantia* var. *charantia/maxima*) × small-fruited (*M. charantia* var. *muricata/minima*) population has indicated that small fruit was partially dominant over large fruit (Kim and Kim 1990). In contrast, fruit length was incompletely dominant and is controlled by a minimum of five genes (Zhang et al. 2006). Likewise, the dramatic role of epistasis in the development of fruits suggests that breeding for this trait will be challenging (Sirohi and Choudhury 1983; Chaudhari and Kale 1991).

Genotypic correlation coefficients in bitter gourd are greater than phenotypic coefficients for most of the traits (Dey et al. 2009). Nevertheless, phenotypic evaluation of yield and quality characteristics used in path coefficient analysis revealed that fruit weight had the greatest direct effect on yield, followed by number of fruits per plant and fruit length. Ascorbic acid content and total carotenoid content had a strong negative but indirect effect on marketable yield based, primarily on fruit weight, fruit length, and diameter. Thus, selection for small-fruited cultivars could improve ascorbic acid and total carotenoid content. Fruit length, average fruit weight, and number of fruits per vine are controlled by additive factors; and thus have direct positive effects on fruit (Sharma and Bhutani 2001). Consequently, simple selection strategies (e.g., backcrossing) focusing on flowering duration, harvesting span, fruit length and diameter, fruit rind thickness, average fruit weight, number of fruits per vine, dry fruit weight, dry matter per vine, and harvest index could be used to improve bitter gourd yield. In contrast, genetic dominance and complementary gene action associated with some of these traits combined with their low narrow-sense heritability indicate that hybrid breeding would be an advantageous strategy when breeding for increased yield in this crop species (Celine and Sirohi 1998; Mishra et al. 1998).

In bitter gourd, nonadditive gene effects were involved for days to first female flowering, fruit weight, fruit flesh weight, fruit girth, number of seeds per fruit, number of fruits per plant, and yield per plant (Khattra et al. 2000; Chowdhury and Sikdar 2005; Islam et al. 2009). The characters like, node to first female flower, vine length, number of female flowers per plant, number of primary branches per plant, average fruit weight, fruit length, number of fruits per vine, average flesh thickness, and yield per plant (Singh and Joshi 1980; Khattra et al. 2000; Sharma and Bhutani 2001; Islam et al. 2009; Dey et al. 2009; Raja et al. 2007) bitter principle (Devdas and Ramdas 1994) were under additive gene control. Yield per plant had high positive and highly significant correlation with number of fruits per plant (Srivastava and Srivastava 1976; Ramachandran and Gopalkrishnan 1979; Kole et al. 2012), number of nodes per vine (Islam et al. 2009), fruit weight, fruit length, and number of flowers per plant (Ramachandran and Gopalkrishnan 1979; Parhi et al. 1995).

The epistasis was important for plant height, fruit length, fruit diameter, fruit weight, and yield/plant. The analysis of variance for sums and differences were significant for all earliness and yield characters, indicating the importance of additive and dominant components of variation for these characters (Ram et al. 2002). The additive as well as nonadditive gene action for fruit yield were reported (Arunachalam 2002).

The ratio between additive and dominance variance revealed the predominance of nonadditive gene action for node to first female flower appearance, days to first female flower appearance, days from fruit setting to maturity, sex ratio, and yield per plant. However, greater role of additive gene effects than dominance was evident for the characters like vine length (Dey et al. 2012). The preponderance of dominant gene action was observed for the traits viz. days to first male and female flower appearance, fruit flesh thickness, number of fruits per vine, yield of fruits per vine, ascorbic acid, and iron content (Thangamani 2016). The additive × dominance and dominance × dominance interaction effects were noted in most of the crosses for the entire traits showed significant epistatic gene effects (Shukla et al. 2014). The characters like days to 50% flowering, days to first harvesting, fruit length, fruit diameter, number of fruits per plant, and yield per plant exhibited predominance of nonadditive gene action, whereas overdominance gene action for most of the yield-related traits were recorded (Rao et al. 2018). The nonadditive gene effect in governing the inheritance of carotene, vitamin C, total sugar, and reducing sugar in bitter gourd was reported by Sundharaiya and Shakila (2011); Kumara et al. (2011) and Kumar and Pathak (2018).

8.2.2 Inheritance for Resistance to Pest and Diseases

Disease and pest cause significant loss in bitter gourd crop and management of these stresses not only increase the cost of production but also has implication on environment and ecology. The increasing use of chemicals for pest, and disease management is concern for growers and human health. The best method to address these stresses is to use resistant varieties which are economical, healthier, and ecofriendly approach. There are different approaches for breeding resistance against different kinds of stresses which involve both conventional and modern tools. Achieving the goal of development of a resistant variety also needs attention on the durability of resistance for which the knowledge of resistance on the inheritance, expression, and interaction with related genes and environment aspects are required.

In bitter gourd, several genetic studies have determined that an association exists between morphological traits and insect resistance and that these associations may be useful for indirect selection during resistance breeding (Dhillon et al. 2005). For instance, percentage of fruit infestation by fruit fly is positively correlated with rib depth, flesh thickness but fruit diameter and length are negatively associated with fruit toughness (Dhillon et al. 2005). Thus, relative fruit toughness might be used as a selection criterion during the development of fruit fly resistant cultivars. In this regard, Tewatia and Dhankhar (1996) reported resistance to fruit fly is dominant, and that additive and dominance gene effects, as well as duplicate epistasis, are important components of resistance. Thus, reciprocal

recurrent selection was suggested as an appropriate breeding strategy for improvement of this trait. Dhillon et al. (2005) observed a significant and positive correlation ($r = 0.96$) between percentage of fruit fly infestation and several fruit characters. In fact, genetic analysis has indicated that total variation for fruit fly infestation and variation for larval density/fruit is associated with variation in flesh thickness and fruit diameter ($r = 0.93$), and flesh thickness and fruit length ($r = 0.76$), respectively. Thus, it appears that phenotypic selection during backcrossing could be practiced directly on these traits for population improvement. Fruit composition components including ascorbic acid, nitrogen, phosphorus, potassium, protein, reducing sugars, nonreducing sugars, and total sugars are negatively correlated with fruit fly resistance, while the moisture content is positively associated with these components. The negative correlation between fruit quality and fruit fly resistance is, in fact, a challenge for breeding programs focused on combining both these traits in improved germplasm. Bitter gourd distortion mosaic virus (BDMV) resistance is controlled by polygenes and their expressions are highly influenced by environment (Arunachalam 2002). Nonadditive gene action for BDMV resistance was reported (Arunachalam 2002). Fruit infestation by this pest has been reported to vary between 41 and 89% in Asia (Rabindranath and Pillai 1986) and up to 95% bitter gourd fruit infestation has been reported in Papua New Guinea (Hollingsworth et al. 1997). Inheritance of resistance to melon fruit fly indicates that fruit fly resistance is dominant over susceptibility (Tewatia and Dhankhar 1996). In an earlier study, Chelliah (1970) also reported that the resistance to melon fruit fly (*D. cucurbitae*) in *Cucumis callosus* appeared to be associated with high silica content of the fruits. It is concluded that higher amounts of tannin, flavonol, total phenol and silica contents of fruits of resistant genotypes may be responsible for imparting resistance against melon fruit fly in bitter gourd. Disease resistance is an important trait in bitter gourd as disease can reduce yield and quality. Powdery mildew caused by *Podosphaera xanthii* is a serious fungal disease of bitter gourd and yield losses of up to 50% have been reported. Resistance was dominated by nuclear genes.

Genetics of several other characters have not been studied and are very limited. More genetic studies are required to know the resistance of morphological characters as well as pest and disease resistance to apply in crop improvement programs.

8.2.3 Molecular Genetic Mapping Efforts

The selection for desirable traits done at the seedling stage overcomes the disadvantages of large field requirements and large number of controlled pollinations. The complication is encountered when selection is required for quantitative traits which are highly influenced by environmental factors. Due to the fact that environmental factors influence quantitative traits, large populations are needed to account for this variability, which adds to time and space requirement and high cost for proper evaluation. It is established that greater number of markers are necessary for the development of a genetic map and MAS strategies (Tang et al. 2007). Complete draft genome sequence of *M. charantia* is still unavailable, and the available DNA markers are limited (Wang et al. 2010; Guo et al. 2012). Therefore, a sequencing-based genotyping method has been employed as a genetic mapping tool in this species (Etter et al. 2011). Elshire et al. (2011) have developed simple and highly multiplexed genotyping by sequencing (GBS) approach for population studies, germplasm characterization, and mapping of desired traits. Reference genome sequences are extremely useful for designing DNA markers such as genome wide single nucleotide polymorphisms (SNPs). Recently, next-generation sequencing (NGS)-based genotyping methods, including Restriction site associated DNA sequencing (RAD-seq; Baird et al. 2008) and GBS (Elshire et al. 2011) have been introduced as genetic mapping tools. These methods are based on sequencing of short fragments from defined

positions in the genome and counting their frequency. Ethylene responsive proteins were identified as putative candidate genes for gynoecy (Gunnaiah et al. 2014). The random amplified polymorphic DNA (RAPD) marker linked to gynoecious (*gy-1*) gene was identified at 22 cM distance (Mishra et al. 2015). The inter-simple sequence repeat (ISSR) marker associated with the gynoecious trait in bitter gourd was amplified by the primer (AC) 8T (Gaikwad et al. 2014). Several SNP loci were identified which are genetically linked to gynoecy in bitter gourd of which the closest SNP markers, GTFL-1 is located at 5.46 cM distance (Matsumura et al. 2014). The first genetic linkage map of bitter melon derived from an inter-botanical variety cross between Taiwan White, *Momordica charantia* var. *charantia*, and CBM12, *M. charantia* var. *muricata* was developed (Kole et al. 2012). Besides, 12 quantitative trait loci (QTLs) controlling five polygenic fruit traits including length, diameter, weight, number, and yield were detected on five linkage groups that individually explained 11.1 to 39.7% of the corresponding total phenotypic variance. Matsumura et al. (2014) identified DNA markers for gynoecy using a RAD-seq analysis. The first genetic map of bitter gourd was constructed based on amplified fragment length polymorphism (AFLP) markers (Kole et al. 2012). A year later, the second genetic map was constructed based on simple sequence repeats (SSR), AFLP, and sequence-related amplified polymorphism (SRAP) markers (Wang and Xiang 2013). An extensive genetic linkage map was constructed using F_2 progenies. The map included 194 loci on 11 chromosomes consisting of 26 expressed sequence tag (EST)-SSR loci, 28 SSR loci, 124 AFLP loci, and 16 SRAP loci. This map covered 1005.9 cM with 12 linkage groups. A total of 43 QTLs, with a single QTL associated with 5.1–33.1% phenotypic variance, were identified on nine chromosomes for 13 horticultural traits. One QTL cluster region was detected on linkage group (LG)-5 which contained the most important QTLs with high contributions to phenotypic variance (5.8–25.4%) (Wang and Xiang 2013). Shukla et al. (2015) performed de novo transcriptome sequencing of bitter gourd using Illumina next-generation sequencer, from root, flower buds, stem and leaf samples of gynoecious line (Gy323) and a monoecious line (DRAR1). The whole transcriptome sequencing of female and hermaphrodite flower buds of bitter gourd was performed using Roche 454 parallel pyrosequencing technology (Behera et al. 2016). Recently, two RAD-based genetic maps were constructed (Matsumura et al. 2014; Urasaki et al. 2017), allowing preliminary attempts at NGS. The draft genome sequence of monoecious bitter gourd line OHB3 was analyzed. Using RAD-seq analysis, a linkage map was constructed, comprising of 11 linkage groups (Urasaki et al. 2017). Cui et al. (2018) reported the construction of a restriction site associated DNA (RAD)-based genetic map for bitter gourd. They also identified three QTL/gene loci responsible for sex expression, fruit epidermal structure, and immature fruit color in three environments. The QTL/gene $gy/fffn/ffn$, controlling sex expression involved in gynoecy, first female flower node, and female flower number was detected. Particularly, two QTLs/genes, Fwa/Wr and w, were found to be responsible for fruit epidermal structure and white immature fruit color, respectively. Rao et al. (2018) developed a high-density, high-resolution genetic map for bitter gourd. A total of 2013 high-quality SNP markers binned to 20 linkage groups (LG) spanning a cumulative distance of 2329.2 cM were developed. The average distance between markers was 1.16 cM across 20 LGs and average distance between the markers ranged from 0.70 (LG-4) to 2.92 cM (LG-20). A total of 22 QTLs for four traits (gynoecy, sex ratio, node, and days at first female flower appearance) were identified and mapped on 20 LGs. The gynoecious ($gy-1$) locus was flanked by markers TP_54865 and TP_54890 on LG 12 at a distance of 3.04 cM to TP_54890. Two major QTLs together with 25.97% phenotypic variance for node to first pistillate flower appearance and both QTLs $qNPF9-F_3$ and $qNPF14-F_2$ showed positive additive effects increased the nodes for first pistillate flower appearance at 1.24 and 0.91 node number, respectively. Similarly, Wang and

Xiang (2013) identified three QTLs ($fffn4.1$, $fff5.1$, and $fffn9.1$) two QTLs in two different locations by Cui et al. (2018) in bitter gourd.

8.2.4 Limitations of Genetic Linkage Mapping

The major limitations of molecular markers are limited in number, along with few relatively economically important traits in bitter gourd are associated with these markers. The full potential of these available molecular maps can only be fully exploited when the entire genome can be visualized on a map and important traits should be associated with the map. The scarcity of polymorphic molecular markers in the public database has hindered genetic mapping and the application of molecular breeding in bitter gourd. The molecular basis of agronomically important traits remains unexplored to date and decisive linkage map has not been reported in bitter gourd (Rao et al. 2018).

8.3 Traditional Breeding

8.3.1 Traditional Breeding Objectives and Achievements

The prime goals of bitter gourd breeding programs are development of high yielding cultivars that possess good quality fruits. Development of cultivars with earliness trait (appearance of first female flower at lower nodes) and high femaleness (high female to male flowers ratio) is another important objective for bitter gourd breeders. To achieve these goals, it differs among the breeders by his/her definition of quality fruits. Disease resistance should be considered as a high priority for most bitter gourd breeding programs in future. Bitter gourds are dynamic plants which have great potential for alteration of quality traits. Quality traits have been selected in bitter gourd recently. These traits include size, shape, shelf life, color, ridgeness, phytochemical, and nutrient content. More recently, breeders have been interested in phytonutrient content, and breeding programs have focused on increasing momordicin, charantin, and iron content in bitter gourd fruits. Although, bitter gourd has a long cultivation history, the molecular research and breeding efforts have been started later than the other major Cucurbitaceae vegetables. Currently, the genetic researches on bitter gourd are mainly focused on yield, quality, maturity period, fruit characteristics, economic traits, etc. Research is also being carried out for disease resistance and application of DNA markers along with traditional breeding. In recent years, various bitter gourd varieties have been bred for earliness, high yield potential, and resistance to diseases using heterosis breeding. The breeding methods practiced in bitter gourd are inbreeding, pure line selection in segregating generations, and heterosis breeding. Selection from a local cultivar has been a most commonly adopted breeding procedure in this crop. So far, very few cultivars have been developed through hybridization. Bitter gourd is a monoecious and highly cross-pollinated crop where a large amount of variation is observed in quantitative and qualitative characters. The exploitation of hybrid vigor in bitter gourd is desirable due to substantial heterosis for earliness and yield traits.

Recently, several studies on heterosis in relation to earliness, yield, and quality traits of bitter gourd have been successfully conducted across the world including ICAR-Indian Agricultural Research Institute, New Delhi, India. Several hybrid and open-pollinated (i.e., usually landraces) cultivars have been released for bitter gourd cultivation (Sirohi 1997), and about 80% of the crop is from established F_1 hybrids. Hybrids usually provide higher yields than open-pollinated cultivars, but hybrid seed is relatively expensive and must be purchased each planting season. In India, the choice of cultivar depends on regional consumer preference for fruit shape, internal and external color, ridging, and degree of bitterness. The first bitter gourd hybrid, Pusa Hybrid-1 was developed and released for commercial cultivation under north Indian plains, which gives 42% heterosis over better parent (Sirohi 2000). The variety Pusa

Hybrid-2 exhibited 75% heterosis for yield/plant over check Pusa Do Mausami. Heterosis has been reported in bitter gourd in desirable directions for several economic traits including earliness (node number of first female flower, days to first female flower appearance), yield and its related traits.

8.3.2 Breeding Methods

Several methods usually are employed in tandem to accomplish breeding objectives. Single-plant selection, mass selection, pedigree selection, and bulk population improvement are common methods used for bitter gourd enhancement (Sirohi 1997). Pedigree selection typically is used after crossing two parents for the development of inbred lines with high, early yield borne on a unique plant habit, and/or with high-quality fruit [i.e., processing quality, high vitamin C and A, and disease resistance]. However, strategies that incorporate selection for disease resistance and improved yield require judicious implementation, since selection for disease resistance can be negatively correlated with yield, as is found in cucumber (Staub and Grumet 1993).

8.3.3 Breeding Systems

The breeding system depends upon the reproduction system of the plant. Information on floral biology is the basic need before setting up a breeding program. There is very little information about the floral biology and genetic system (number of genes and chromosomes, details of meiosis and pairing, breeding system, sex determination and sex modification, and regulation of gene actions) in *Momordica* species except for bitter gourd (*M. charantia*) and to some extent in *M. dioica*.

8.3.3.1 Heterosis Breeding
As all the species of *Momordica* are cross-pollinated, there is ample scope for exploitation of heterosis. More pronounced hybrid vigor

could be observed with the inclusion of diverse parents. Heterosis for earliness, high number of fruits, and bearing at each flowering node should be exploited. Selection for divergent parent based on number of fruits, fruit weight, fruit length, internodal length, pedicel length, and yield will be useful as these characters were the major traits contributing to divergence in *M. dioica* (Bharathi et al. 2005). Heterosis in *M. charantia* was investigated at the Indian Agricultural Research Institute, New Delhi, as early as 1943 (Pal and Singh 1946). Evidence of heterotic effects is supported by genetic analyses that have defined the presence of dominance and complementary gene action for yield as distinguished by its components (Mishra et al. 1998). Heterosis for yield per vine ranges from 27 to 86% depending on genotype (Behera 2004). This heterosis is likely attributable to earliness, first node to bear fruit (first female flowering node), and total increased fruit number (Celine and Sirohi 1998). A few hybrids have been developed at both private and public sectors for cultivation in Asia including China and India. Appreciably high amount of significant heterosis was observed in desirable direction for node number of first female flower, days to first female flower appearance, number of fruits per plant, average fruit weight, fruit length, fruit diameter, vine length, number of primary branches (Lawande and Patil 1989; Jadhav et al. 2009; Bhatt et al. 2017), node at which first female flower appeared and fruit length (Thangmani and Pugalendhi 2013),days to first harvest, fruit length, fruit diameter, fruit weight, number of fruit/plant, and fruit yield/plant (Pal and Singh 1946; Lawande and Patil 1989; Singh and Joshi 1980). The F_1 hybrids had reportedly high yield potential over their parents in bitter gourd (Behera et al. 2005; Thangamani and Pugalendhi 2013; Singh et al. 2013; Bhatt et al. 2017). For the first time, bitter gourd plants with a complete expression of gynoecious flowering habit have been located in naturally occurring three independent lines (Ram et al. 2002). Two gynoecious lines (DBGy-201, DBGy-202) were isolated from indigenous

source, from its related wild form, M. *charantia* var. *muricata* L. (Behera et al. 2006). Another two gynoecious lines, (IIHRBTGy-491 and IIHRBTGy-492), have also been identified in bitter gourd (Varalakshmi et al. 2014). The F_1 hybrids developed using gynoecious lines were found superior in performance for several earliness and yield characters (Dey et al. 2012; Alhariri et al. 2018; Rao et al. 2018). The hybrids showed heterosis in desirable direction for sex ratio, days to opening of first female flower and days to first picking. The monoecy-based hybrids were superior for fruit length, fruit diameter, flesh thickness, average fruit weight and yield per plant (Alhariri et al. 2018; Rao et al. 2018). Positive heterosis was reported over better parent for vitamin C, carotene and total sugar content in bitter gourd (Adarsh et al 2018).

8.3.3.2 Polyploidy Breeding

Polyploids can be produced in bitter gourd by treating the seedlings at the cotyledon stage with an emulsion of 0.2% colchicine. Yasuhiro Cho et al. (2006) reported that seed treatment with 0.2 and 0.4% colchicine or 0.003% amiprophos-methyl was effective for chromosome doubling, among which the treatment with 0.4% colchicine was most effective. Amiprophos-methyl treatment also resulted in octaploid plants with high rate of seed germination. Multiple shoot treatments with 0.05% colchicine for 12 and 24 h, and 0.1% colchicine for 24 h also led to octaploid plants. Leaf and guard cell size were bigger, and leaf shape index (leaf length/leaf width) was lower in the octaploid than in tetraploid plants. Leaves of the octaploid plants were uneven on the surface with clear serrations. Triploid plants of *M. charantia* were obtained by crossing the tetraploid (colchicines induced) and diploid plants (Saito 1957). In seedlings of bittergourd, colchicine at 0.2% for 18 h to the shoot tip produced tetraploids (Kadir and Zahoor 1965). However, polyploids were inferior to diploids in terms of economic characters. Reyes and Rasco (1994) reported a suppressed shoot growth mutant due to the recessive gene *ssg*.

8.4 Limitations of Traditional Breeding and Rationale for Molecular Breeding

Traditional breeding has relied, either directly or indirectly, on morphological markers to identify the trait of interest for selection in a segregating population. Traditional breeding includes a direct measure of phenotypes (e.g., sex ratio, fruit color) or an association of one phenotype with another (dark green leaves with bitter fruits). Traditional breeding has been extremely effective for making qualitative changes in bitter gourd crop. In addition to improvement in qualitative traits, there have been huge changes in important quantitative traits through traditional breeding. Further improvement in quantitative traits using traditional strategies will be time consuming. Breeding for disease resistance is often challenging in bitter gourd crop. Many disease resistance traits are quantitative, there expression is often affected by environment and therefore, a complex inoculation procedure is required.

Molecular markers have the potential for overcoming the limitations associated with traditional breeding, since they are nondestructive, can eliminate the environmental variation associated with a trait (viz. disease resistance) and can be evaluated for multiple traits simultaneously. However, the use of molecular markers requires the development of a mapping population, which segregates for the trait of interest, and the trait must be properly identified for marker-assisted selection. There are two important issues to consider regarding molecular markers: time and cost. Although the molecular breeding and the development of molecular markers have great potential, it may take significant amount of time for proper development of the marker and thorough test of the marker in multiple populations. The potential for molecular markers to save money is in their long-term utilization in combination with multiple markers for a wide variety of traits; this will allow breeders to select for multiple traits from large populations which is currently not possible.

8.5 Concluding Remarks and Future Prospects

Classical genetics and traditional breeding have made enormous strides in understanding and improvement of bitter gourd crop. We have successfully used classical genetics to enhance our understanding of taxonomy and phylogenetic relationship in bitter gourd. Bitter gourd breeders have identified a number of genes associated with economically important traits and have used this information to create superior cultivars with high yield, earliness, and quality. The previous advances were typically through adding single traits with high heritability to adapted germplasm to a particular environment. Advancement of some quantitative and qualitative traits using traditional techniques is difficult and also time-consuming process since the complexity increases with each added traits. Molecular breeding plays significant role to overcome many of the problems associated with traditional breeding and genetics. As we move forward with molecular breeding in bitter gourd, it is important that we understand the need to maintain traditional breeding programs, and that the skill set required for classical breeding is not lost. Developing molecular markers requires populations with traits identified using traditional methods, at least in the initial stages. If we want to fully exploit the potential of molecular breeding, it is utmost important that we maintain the balance between molecular marker technology and traditional breeding. This chapter delves into the advances made in bitter gourd breeding regardless of the breeding method used.

References

Adarsh A, Kumar R, Singh HK, Bhardwaj A (2018) Heterosis study in bitter gourd for earliness and qualitative traits. Intl J Curr Microbiol Appl Sci 7:4239–4245

Ahmed I, Lakhani MS, Gillett M, John A, Raza H (2001) Hypotriglyceridemic and hypocholesterolemic effects of anti-diabetic *Momordicacharantia* (karela) fruit extract in streptozotocin-induced diabetic rats. Diab

Res Clin Pract 51:155–161. https://doi.org/10.1016/s0168-8227(00)00224-2

Alhariri A, Behera TK, Munshi AD, Bharadwaj C, Jat GS (2018) Exploiting gynoecious line for earliness and yield traits in bitter gourd (*Momoredicacharantia* L.). Int J Curr Microbiol Appl Sci 7(11):922–928

Arunachalam P (2002) Breeding for resistance to distortion mosaic virus in bitter gourd (*Momordica charantia* L.). Ph.D. thesis, Faculty of Agriculture, Kerala Agriculture University, Kerala, India

Baird NA, Etter PD, Atwood TS, Currey MC, Shiver AL et al (2008) Rapid SNP discovery and genetic mapping using sequenced RAD markers. PLoS ONE 3:e3376. https://doi.org/10.1371/journal.pone.0003376

Bates DM, Merrick LC, Robinson RW (1995) Minor cucurbits. In: Smartt J, Simmonds NW (eds) Evolution of crop plants. Wiley, New York, p 110

Behera TK (2005) Heterosis in bittergourd. J New Seeds 6 (2):217–221

Behera TK, Behera S, Bharathi LK, John KJ, Simon PW, Staub JE (2010) Bitter gourd: botany, horticulture and breeding. Hort Rev 37:101–141

Behera TK, Dey SS, Munshi AD, Gaikwad AB, Pal A, Singh I (2009) Sex inheritance and development of gynoecious hybrids in bitter gourd (*Momordicacharantia* L.). Sci Hort 120:130–133

Behera TK, Rao AR, Amarnath R, Kumar RR (2016) Comparative transcriptome analysis of female and hermaphrodite flower buds in bitter gourd (*Momordica charantia* L.) by RNA sequencing. J Hort Sci Biotechnol 91(3):250–257. https://doi.org/10.1080/14620316.2016.1160540

Behera TK, Dey SS, Sirohi PS (2006) DBGy-201and DBGy-202: two gynoecious lines in bitter gourd (*Momordica charantia* L.) isolated from indigenous source. Indian J Genet 66:61–62

Behera TK (2004) Heterosis in bittergourd. In: Singh PK, Dasgupta SK, Thpathi SK (eds) Hybrid vegetable development. Haworth Press, New York, pp 217–221

Bharathi LK, Naik G, Dora DK (2005) Genetic divergence in spine gourd. Veg Sci 32(2):179–81

Bhatt L, Singh SP, Soni AK, Samota MK (2017) Combining ability studies in bitter gourd (*Momordica charantia* L.) for quantitative characters. Int J Curr Microbiol Appl Sci 6(7):4471–4478

Celine VA, Sirohi PS (1998) Generation mean analysis for earliness and yield in bitter gourd (*Momordica charantia* L.). Veg Sci 25:51–54

Chaudhari SM, Kale PN (1991) Studios on heterosis in bitter gourd (*Momordica charantia* L.). Maha J Hort 5:45–51

Chelliah S (1970) Host influence on the development of melon fly (*Dacus cucurbitae* Coquillett). Indian J Entomol 32:381–383

Chen Q, Chan LL, Li ET (2003) Bitter melon (*Momordica charantia*) reduces adiposity lowers serum insulin and normalizes glucose tolerance in rats fed a high fat diet. J Nutr 133:1088–1093. https://doi.org/10.1093/jn/133.4.1088

Cho Y, Ozaki Y, Okubo H, Matsuda S (2006) Ploidies of kakrol (*Momordica dioca* Roxb.) cultivated in Bangaladesh. Sci Bull Fac Agri Kyushu Univ 61:49–53

Chowdhury AR, Sikdar B (2005) Genetic analysis for nine fruit characters in relation to five parental diallel crossing of bitter gourd. J Life Earth Sci 1:31–34

Cui J, Luo S, Niu Y, Huang R, Wen Q, Su J, Miao N, He W, Dong Z, Cheng J, Hu K (2018) A RAD-based genetic map for anchoring scaffold sequences and identifying QTLs in bitter gourd (*Momordica charantia*). Front Plant Sci 9:477. https://doi.org/10.3389/fpls.2018.00477

Dalamu, Behera TK, Satyavati TC, Pal A (2012) Generation mean analysis of yield related traits and inheritance of fruit colour and surface in bitter gourd. Indian J Hort 69:65–69

Devadas VS, Ramadas S (1994) Combining ability for yield components in bitter gourd (*Momordica charantia* L.) Hort J 6:103–108

Dey SS, Behera TK, Munshi AD, Pal A (2009) Gynoecious inbred with better combining ability improves yield and earliness in bitter gourd (*Momordica charantia* L.). Euphytica 173(1):37–47

Dey SS, Behera TK, Munshi AD, Rakshit S, Bhatia R (2012) Utility of gynoecious sex form in heterosis breeding of bitter gourd and genetics of associated vegetative and flowering traits. Indian J Hort 69 (4):523–529

Dhillon MK, Singh R, Naresh JS, Sharma NK (2005) Influence of physico-chemical traits of bitter gourd, *Momordica chamntia* L. on lanai density and resistance to melon-fruit fly, *Boctrocem cucurbitae* (Coquillett). J Appl Entomol 129:393–399

Elshire RJ, Glaubitz JC, Sun Q, Poland JA, Kawamoto K et al (2011) A robust, simple genotyping-by-sequencing (GBS) approach for high diversity species. PLoS ONE 6:e19379. https://doi.org/10.1371/journal.pone.0019379

Esquinas-Alcazar JT, Gulick PJ (1983) Genetic resources of Cucurbitaceae. AGPGR: IBPGR/83/48:20

Etter PD, Bassham S, Hohenlohe PA, Johnson EA, Cresko WA (2011) SNP discovery and genotyping for evolutionary genetics using RAD sequencing. Meth Mol Biol 772:157–178. https://doi.org/10.1007/978-1-61779-228-1_9

Gaikwad AB, Saxena S, Behera TK, Archak S, Meshram SU (2014). Molecular marker to identify gynoecious lines in bitter gourd. Indian J Horticulture 71(1):142–144

Ganguly C, De S, Das S (2000) Prevention of carcinogen induced mouse skin papilloma by whole fruit aqueous extract of *Momordica charantia*. Eur J Cancer Prev 9:283–288. https://doi.org/10.1097/00008469-200008000-00009

Gunnaiah R, Vinod MS, Prasad K, Elangovan M (2014) Identification of candidate genes, governing gynoecy in bitter gourd (Momordica Charantia L.). In: Silico gene expression analysis. National conference cum workshop on bioinformatics and computational biology, NCWBCB- 2014. Int J Comput Appl 5–9

Guo DL, Zhang JP, Xue YM, Hou XG (2012) Isolation and characterization of 10 SSR markers of *Momordicacharantia* (Cucurbitaceae). Am J Bot 99:e182–e183. https://doi.org/10.3732/ajb.1100277

Hollingsworth R, Vagalo M, Tsatsia F (1997) Biology of melon fly with special reference to the Solomon Islands. In: Allwood AJ, Drew RAI (eds) Management of fruit flies in the Pacific (Proc Aust Country Ind Agri Res 76:140–144)

Hossain MA, Islam M, Ali M (1996) Sexual crossing between two genetically female plants and sex genetics of kakrol (*Momordica dioica* Roxb.) Euphytica 90 (1):121–125

Huang HY, Hsieh CH (2017) Genetic research on fruit color traits of the bitter gourd (*Momordica charantia* L.). Hort J. https://doi.org/10.2503/hortj.mi-159

Hu KL, Fu QM, Wang GP (2002) Study on the heredity of fruit color of *Monordica charantia*. China Veg 2002:11–12

Islam MR, Hossain MS, Buiyan MSR, Husna A, Syed MA (2009) Genetic variability and path coefficient analysis of bitter gourd (*Momordica charantia* L.). Int J Sustain Agri 1(3):53–57

Iwamoto B, Ishida T (2006) Development of gynoecious inbred line in balsam pear (*Momordica charantia* L.). Hort Res (Japan) 5:101–104

Jadhav KA, Garad BV, Dhumal SS, Kshirsagar DB, Patil BT Shinde KG (2009) Heterosis in bitter Gourd (*Momordica charantia* L.). Agri Sci Digest 29(1):7–11

Jayasooriya AP, Sakono M, Yukizaki C, Kawano M, Yamamoto K, Fukuda N (2000) Effects of *Momordicacharantia* powder on serum glucose levels and various lipid parameters in rats fed with cholesterol-free and cholesterol enriched diets. J Ethnopharmacol 72:331–336

Kadir ZBA, Zahoor M (1965) Colchiploidy in *Momordicacharantia* L. Sind Uni Res J 1–53

Khattra AS, Singh R, Thakur JC (2000) Combining ability studies in bitter gourd in relation to line× tester crossing system. Veg Sci 27:148–151

Kim ZH, Kim YR (1990) Inheritance of fruit weight in bitter-gourd (*Mornordica charantia* L.). J Korean Soc Hort Sci 31:238–246

Kole C, Olukolu BA, Kole P, Rao VK, Bajpai A, Backiyarani S, Singh J, Elanchezhian R, Abbott Albert G (2012) The first genetic map and positions of major fruit trait loci of bitter melon (*Momordica charantia*). J Plant Sci Mol Breed https://doi.org/10.7243/2050-2389-1-1

Krishnaiah D, Sarbatly R, Nithyanandam R (2011) A review of the antioxidant potential of medicinal plant species. Food Bioprod Process 89:217–233

Kumar D, Pathak M (2018) Estimation of heterosis and combining ability for biochemical traits in bitter gourd (*Momordicacharantia* L.). Int J Chem Stud 6 (2):2579–2585

Kumara BS, Puttaraju TB, Hongal S, Prakash K, Jainag K, Sudheesh NK et al (2011) Combining ability studies in bitter gourd (*Momordica charantia* L.) for quantitative characters. Asian J Hort 6:135–140

Kumari M, Behera TK, Munshi AD, Talukadar A (2015) Inheritance of fruit traits and generation mean analysis for estimation of horticultural traits in bitter gourd. Indian J Hort 72(1):43–48

Lawande KE, Patil AV (1989) Studies on heterosis as influenced by combining ability in bitter gourd. Veg Sci 6:49

Lee HS, Huang PL, Huang PL, Bourinbaiar AS, Chen HC, Kung HF (1995) Inhibition of the integrase of human immuno-deficiency virus (HIV) type 1 by anti-HIV plant proteins MAP30 and GAP31. Proc Natl Acad Sci USA 92:8818–8822. https://doi.org/10.1073/pnas.92.19.8818

Leung KC, Meng ZQ, Ho WKK (1997) Antigenic determination fragments of α-momorcharin. Biochem Biophys Acta 1336:419–424

Liou TD, Chen KS, Lee SF, Lin JNT, Sao SJ, Yang YW (2002) Tangshan 036', a white bitter gourd cultivar. HortScience 37:1142–1143

Liu ZG, Long MM, Qin RY, Wang XY (2005) Studies on genetic variation, correlation and path analysis in bitter gourd (*Mornordica charantia* L.). Guan Bot 25:426–430

Matsumura H, Miyagi N, Taniai N, Fukushima M, Tarora K et al (2014) Mapping of the gynoecy in bitter gourd (*Momordica charantia*) using RAD-Seq analysis. PLoS ONE 9(1):e87138. https://doi.org/10.1371/journal.pone.0087138

Miniraj N, Prasanna KP, Peter KV (1993) Bitter gourd (*Momordica* spp.). In: Kalloo C, Bergh BO (eds) Genetic improvement of vegetable plants. Pergamon Press, Oxford, UK, pp 239–246

Mishra S, Behera TK, Munshi AD, Gaikwad K, Mohapatra T (2015) Identification of RAPD marker associated with gynoecious trait (*gy-1*gene) in bitter gourd (*Momordica charantia* L.) Aust J Crop Sci 8(5):706–710

Mishra HN, Mishra RS, Parhi G, Mishra NS (1998) Diallel analysis for variability in bitter gourd (*Momordica charantia* L.). Indian J Agri Sci 68:18–20

Myojin C, Enami N, Nagata A, Yamaguchi T, Takamura H, Matoba T (2008) Changes in the radical-scavenging activity of bitter gourd (*Momordica charantia* L.) during freezing and frozen storage with or without blanching. J Food Sci 73:46–50

Nicoli MC, Anese M, Parpinel M, Kubola J, Siriamornpun S (2008) Phenolic contents and antioxidant activities of bitter gourd (*Momordica charantia* L.) leaf, stem and fruit extracts in vitro. Food Chem 110:881–890

Okabe H, Miyahara Y, Yamauchi T, Mirahara K, Kawasaki T (1980) Studies on the constituents of *Momordica charantia* L. Isolation and characterization of momordicosides A and B, glycosides of a pentahydroxy-cucurbitanetriterpene. Chem Pharm Bull 28:2753–2762

Pal HP, Singh H (1946) Studies in hybrid vigour. II. Notes on the manifestation of hybrid vigour in the brinjal and bitter gourd. Indian J Genet 6:19–33

Panda A, Singh DK, Bairagi SK (2008) Study of combining ability for the production of bitter gourd hybrids. Prog Hort 40:33–37

Parhi G, Mishra HN, Mishra RS (1995) Correlation and path coefficient studies in bitter gourd. Indian J Hort 52(2):132–136

Rabindranath K, Pillai KS (1986) Control of fruit fly of bitter gourd using synthetic pyrethroids. Entomon 11:269–272

Raja S, Bagle BG, Dhandar DG (2007) Genetic variability studies in bitter gourd for zero irrigated condition of semi-arid ecosystem. Indian J Hort 64(4):425–429

Ram D, Kumar S, Singh M, Rai M, Kalloo G (2006) Inheritance of gynoecism in bitter gourd (*Momordica charantia* L.). J Hered 97:294–295

Ram D, Kumar S, Banerjee MK, Kellen G (2002) Occurrence, identification and preliminary characterisation of gynoecism in bitter gourd (*Momordica charantia* L.). Indian J Agri Sci 72:348–349

Ramachandran C, Gopalkrishnan PK (1979) Correlation and regression studies in bitter gourd. Indian J Agri Sci 49(11):850–854

Rao GP, Behera TK, Gaikwad AB, Munshi AD, Jat GS, Krishnan B (2018) Mapping and QTL analysis of gynoecy and earliness in bitter gourd (*Momordica charantia* L.) using genotyping-by-sequencing (GBS) technology. Front Plant Sci 9:1555. https://doi.org/10.3389/fpls.2018.01555

Reyes MEC, Rasco ET (1994) Induction and inheritance of restricted vine growth mutant in bittergourd (*Momordica charantia* L.). University Library, University of the Philippines at Los Baños PÁGINA DE INICIO: http://www.uplb.edu.ph

Robinson RW, Decker-Walters DS (1997) Cucurbits. CABI Publishing, Cambridge, MA

Saito K (1957) Studies on the induction of polyploidy in some cucurbits and its utilization: II. On polyploid plants of bitter gourd. Jpn J Breed 6:217–220

Sanwal SK, Kozak M, Kumar S, Singh B, Deka BC (2011) Yield improvement through female homosexual hybrids and sex genetics of sweet gourd (*Momordicacochinchinensis*Spreng.) Acta Physiol Plant 33(5):1991–1996

Sharma NK, Bhutani RD (2001) Correlation and path analysis studies in bitter gourd (*Momordica charantia*). Hary J Hort Sci 30:84–86

Shukla A, Singh U, Rai AK, Bhardwaj DR, Singh M (2014) Genetic analysis of yield and yield attributing traits in bitter gourd. Veg Sci 41:37–41

Shukla A, Singh VK, Bharadwaj DR, Kumar R, Rai A, Rai AK et al (2015) De Novo assembly of bitter gourd transcriptomes: gene expression and sequence variations in gynoecious and monoeciouslines. PLoS ONE 10(6):e0128331. https://doi.org/10.1371/journal.pone.0128331

Singh AK, Pan RS, Bhavana P (2013) Heterosis and combining ability analysis in bitter gourd (*Momordica charantia* L.). Bioscan 8:1533–1536

Singh B, Joshi S (1980) Heterosis and combining ability in bitter gourd. Indian J Agri Sci 50:558–561

Singh SK, Ram HH (2005) Seed quality attributes in bitter gourd (*Momordica charantia* L.). Seed Res 33:92–95

Sirohi PS (1997) Improvement in cucurbit vegetables. Indian Hort 42:64–67

Sirohi PS (2000) Pusa hybrid 1: new bitter gourd hybrid. Indian Hort 44:30–31

Sirohi PS, Choudhury B (1983) Diallelanalysis of variability in bitter gourd. Indian J Agri Sci 53:880–888

Srivastava VK, Nath P (1972) Inheritance of some qualitative characters in *Momordica charantia* L. Indian J Hort 29:319–321

Srivastava VK, Srivastava LC (1976) Genetic parameter correlation coefficient and path coefficient analysis in bitter gourd (Momordica charantia L.). Indian J Horticulture 33:66–70

Staub JE, Grumet R (1993) Selection for multiple disease resistance reduces cucumber yield potential. Euphytica 67:205–213

Sundharaiya K, Shakila A (2011) Line x tester analysis in bitter gourd (*Momordicacharantia* L.). Adv Plant Sci 24:637–641

Suribabu B, Reddy NE, Ramarao M (1986) Inheritance of certain quantitative and qualitative characters in bitter gourd (*Momordica charantia* L.). South Indian Hort 34:380–382

Tang J, Leunissen JA, Voorrips RE, van der Linden CG, Vosman B (2008) HaploSNPer: a web-based allele and SNP detection tool. BMC Genet. 9:23

Tewatia AS, Dhankar B S (1996) Inheritance of resistance to melon fruit fly (Bactrocera cucurbitae) in bitter gourd (Momordica charantia L.). Indian J Agr Sci 66:617–620

Tewatia AS, Dhankhar BS, Dhankhar SK (1997) Growth and yield characteristics of melon fruit fly resistant and highly susceptible genotypes of bitter gourd—a note. Hary J Hort Sci 25:253–255

Thangamani C (2016) Genetic analysis in bitter gourd (*Momordica charantia* L.) for yield and component characters. Asian J Hort 11(2):313–318

Thangamani C, Pugalendhi L (2013) Heterosis studies in bitter gourd for yield and related characters. Int J Veg Sci 19:109–125

Urasaki N, Takagi H, Natsume S, Uemura A, Taniai N, Miyagi N, Fukushima M, Suzuki S, Tarora K, Tamaki M, Sakamoto M, Terauchi R, Matsumura H (2017) Draft genome sequence of bitter gourd (*Momordica charantia*), a vegetable and medicinal plant in tropical and subtropical regions. DNA Res 24 (1):51–58

Vahab MA (1989) Homeostatic analysis of components of genetic variance and inheritance of fruit colour, fruit shape, and bitterness in bitter gourd (*Momordica charantia* L.). Ph.D. Thesis, Kerala Agri Univ, India

Varalakshmi B, Pitchaimuthu M, Rao ES, Krishnamurthy D, Suchitha Y, Manjunath KSS (2014) Identification preliminary characterization and maintenance of gynoecious plants, IIHRBTGy-491 and IIHRBTGy-492 in bitter gourd. Int Bitter Gourd Conf (BiG2014). Hyderabad, March, AVRDC at ICRISAT, p 36

Wang SZ, Pan L, Hu K, Chen CY, Ding Y (2010) Development and characterization of polymorphic microsatellite markers in *Momordica charantia* (Cucurbitaceae). Am J Bot 97:e75–e78. https://doi.org/10.3732/ajb.1000153.16

Wang Z, Xiang C (2013) Genetic mapping of QTLs for horticulture traits in a F2-3 population of bitter gourd (Momordica charantia L.). Euphytica 193:235–250

Yeh GY, Eisenber DM, Kaptchuk TJ, Phillips RS (2003) Systematic review of herbs and dietary supplements for glycemic control in diabetes. Diab Care 26:1277–1294

Yuan J, Njiti VN, Meksem K, Iqbal MJ, Triwitayakorn K, Kassem MA et al (2002) Quantitative trait loci in two soybean recombinant inbred line populations segregating for yield and disease resistance. Crop Sci 42:271–277

Zhang C, Luo S, Goo J, Zheng X, Lao H, Man J (2006) Study on the genetic effects of fruit length of bitter gourd. Guan Agri Sci 1:34–35

Molecular Linkage Mapping in Bitter Gourd

9

Hideo Matsumura, Naoya Urasaki and Chittaranjan Kole

Abstract

Genetic mapping of genes controlling agronomic traits and various phenotypes including the content of phytomedicines is essential for identifying their corresponding genes and linked markers for genetic improvement in bitter gourd. As studied in other crops, various molecular markers have been employed for genotyping including RAPD, AFLP, SSR, ISSR, etc. Recently, next-generation sequencing technology allowed to find a large number of DNA polymorphisms as molecular markers at a time. Sequencing-based molecular marker development, like RAD-seq or GBS, is equally applicable to even nonmodel plants as bitter gourd. Using molecular markers, several linkage maps were independently developed in bitter gourd. Although the number of analyzed markers and linkage groups was different among these maps, these maps were employed for QTL mappings. Currently identified QTLs in bitter gourd were mainly focused on the characteristics for flowers or fruits. These traits determine the quality or yield of the bitter gourd fruit. According to genetic mapping results, fruit color, fruit glossiness, fruit surface structure, stigma color, seed color, and gynoecy were suggested to be qualitative traits and determined by single or double loci. On the other hand, sex ratio, first node of female flower, days to first female flower, several traits for fruits (size, length, etc.) were found to be determined by several QTLs. Due to different mapping populations and employed molecular markers, correspondence among independently identified QTLs were still undetermined, but reference genome sequences of bitter gourd will be helpful for identifying corresponding genes for these QTLs in future.

9.1 Introduction

It is highly important to locate genes controlling various economic traits or phenotypes for genetic improvement of any crop (Kole and Gupta 2004). Since marker-assisted selection (MAS) is useful in breeding of crops, genetic mapping of agronomically important traits directly contribute to the breeding programs (Kole and Abbott 2008). Recent progress of next-generation sequencing (NGS) technology facilitated

H. Matsumura (✉)
Gene Research Center, Shinshu University, Ueda, Nagano, Japan
e-mail: hideoma@shinshu-u.ac.jp

N. Urasaki
Okinawa Agricultural Research Center, Itoman, Okinawa, Japan

C. Kole
ICAR-National Institute for Plant Biotechnology, Pusa, New Delhi, India

© Springer Nature Switzerland AG 2020
C. Kole et al. (eds.), *The Bitter Gourd Genome*, Compendium of Plant Genomes,
https://doi.org/10.1007/978-3-030-15062-4_9

discovery of numerous polymorphisms as molecular markers required for linkage map construction and mapping of simply inherited and quantitative trait loci (QTLs). In this chapter, current advances in detection of molecular markers and their use in linkage map development in bitter gourd were enumerated, and several examples of genetic mapping of genes and QTL in bitter gourd have been presented.

9.2 Molecular Markers in Bitter Gourd

Molecular markers, based on DNA polymorphisms, are essential for linkage map development and genetic mapping of genes and QTLs. In bitter gourd, several genotyping or genetic mapping studies using molecular markers (DNA marker) are reported, although published works are limited. Kole et al. (2012) developed the first genetic map in bitter gourd by using amplified fragment length polymorphism (AFLP) markers. Subsequently, Wang and Xiang (2013) reported a genetic linkage map by utilizing several kinds of markers, including simple sequence repeat (SSR), AFLP, and sequence-related amplified polymorphism (SRAP). In these markers, DNA fragment length or patterns were discriminated in electrophoresis for detection of polymorphism. Since reference genome sequence is unnecessary for developing these markers, they were widely employed in genetic mapping studies. However, they required screening of markers showing polymorphisms between the parental lines, and then genotyping of a segregating population such as F_2 using selected markers. Therefore, the number of applicable markers for mapping was limited.

9.2.1 Sequencing-Based Markers

Now, owing to the recent progress of NGS technology, whole genome sequences were extensively studied in various species and a large number of sequence polymorphisms were also more rapidly found even in nonmodel life organisms. Particularly, sequencing-based genotyping methods like RAD-seq (Baird et al. 2008) or genotyping by sequencing (GBS, Elshire et al. 2011) was quite effective for nonmodel plant species. In this method, genomic DNA fragments digested with any restriction enzymes were prepared and ends of these fragments were massively sequenced. By comparing these sequence reads between parental lines or reference mapping, any kinds of polymorphisms were identified. In these analyses, although only less than 10% regions in the whole genome could be sequenced, hundreds to thousands of polymorphic loci could be found at the same time. Since polymorphisms are detected based on DNA sequence in these methods, genotypes could be digitally determined. The same experimental procedure as that in the parental lines was equally applicable to F_2 individuals for genotyping polymorphic loci. Since indexing or barcoding method was routinely available in NGS library preparation, sequencing of hundreds of individual F_2 RAD-seq or GBS libraries allow to be completed in a single sequence run.

In the RAD-seq analysis of bitter gourd by Matsumura et al. (2014), it was suggested that choice of restriction enzymes was influential, because GC content of genome sequence in higher plants was not so high, and fidelity and methylation sensitivity of employed enzymes should be carefully considered for avoiding inconsistency of genotyping. In this study, codominant markers were defined by selecting either parental line-specific RAD tags or by finding putative bi-allelic tags by sequence comparison of either parent-specific tags. Similarly, RAD-seq analysis using six-base cutter *Ase*I was carried out in two parental lines (Urasaki et al. 2017), but bi-allelic RAD tags were selected from uniquely mapped tags to the bitter gourd reference genome. It allowed to extract only the tags corresponding to the alleles and eliminate paralogous tag sequences. Similarly, sequence reads from *Eco*RI-based RAD-seq analysis of parental lines were aligned against scaffold from whole genome sequencing analysis for identifying single-nucleotide polymorphisms (SNPs) by Cui et al. (2018). Recently,

Gangadhara Rao et al. (2018) employed GBS technique for identifying molecular markers in bitter gourd. In this study, SNP markers were identified without reference mapping of sequencing reads. Although number of identified markers in these analyses depended on the employed restriction enzymes, it was not always extensively increased in GBS (Gangadhara Rao et al. 2018) even when *Apek*I (5'-GGWCC-3') was used compared to RAD-seq using six-base cutters, *Ase*I or *Eco*RI (Urasaki et al. 2017; Cui et al. 2018). This is probable that sequence divergences among bitter gourd lines were low indicating that the available number of markers for mapping might be limited.

9.3 Construction of Genetic Linkage Maps

The first genetic linkage map in bitter gourd was developed by genotyping 146 F_2 individuals from Taiwan White x CBM12 using 108 AFLP markers (Kole et al. 2012). They obtained 11 linkage groups spanning a total of 3060.7 cM with an average marker interval of 22.75 cM, where some groups were comprised of only a few markers (Table 9.1). However, using this map, it was possible to detect five loci controlling five simply inherited traits including fruit color, fruit luster, fruit surface structure, stigma color and seed color, and quantitative trait loci (QTLs) underlying five polygenic traits loci including fruit yield and its four component traits such as length, diameter, weight, and number (Fig. 9.1). Wang and Xiang (2013) constructed another linkage map, based on genotyping 144 F_2 progenies of Z-1-4 x 189-4-1 by 194 loci of several types of markers. In total, 12 linkage groups were developed in their map, while there seemed to be biased in marker distribution. QTLs were also determined using these markers and linkage map.

As described above, traditional DNA markers were conventional but there was a problem in marker density (the number of markers). It is advantageous that RAD-seq or GBS, as sequencing-based genotyping, allowed to find more markers. As described above, it was applicable to both exploration of polymorphisms (markers) in parental lines and marker genotyping in segregating population. In bitter gourd, RAD-seq or GBS was used to genotype about a hundred of F_2 or $F_{2,3}$ individuals for developing linkage maps (Urasaki et al. 2017; Cui et al. 2018; Gangadhara Rao et al. 2018). Generally, a number of accurate molecular markers and segregating populations like F_2 individuals were the key factors for high-density linkage map development. Two linkage maps, developed by RAD-seq analysis by Urasaki et al. (2017) and Cui et al. (2018), showed the number of linkage groups identical to the chromosome numbers (11 chromosomes) in bitter gourd. In these studies, RAD-seq tags or sequence reads were mapped to the genome sequence for calling SNPs or indels. This process avoids sequence differences among redundant regions, which were not derived from alleles. On the other hand, there was a strategy for genotyping by RAD-seq or GBS without reference genome and linkage maps were actually developed in bitter gourd (Matsumura et al. 2014; Gangadhara Rao et al. 2018). However, de novo GBS-based linkage maps were separated in more linkage groups than chromosomes.

Several linkage maps were known in bitter gourd as described above. Still, integration of these linkage maps was not yet done, since mapping populations were independent and most markers were not anchored to the publicly available reference genome sequences except for the RAD-seq markers in Urasaki et al. (2017). SNPs or single-nucleotide variants (SNVs) loci could be easily allocated to genome sequence, since their sequences were known. Correspondence between the linkage map and genome sequences was necessary for identifying genes for genetically mapped traits and constructing pseudomolecules of bitter gourd genome.

9.4 Genetic Mapping of Qualitative and Quantitative Traits

In the reported linkage maps of bitter gourd, DNA markers seemed to be distributed in each linkage group, indicating that these maps and

Table 9.1 Summary of current published linkage map in bitter gourd

Number of linkage groups	Number of marker loci	Marker type	Mapping population	Literature
11	113	AFLP	146 F_2 (Taiwan White x CBM12)	Kole et al. (2012)
12	194	SSR,AFLP, SRAP	144 F_2 (Z-1-4 x 189-4-1)	Wang and Xiang (2013)
11	1507	RAD	97 F_2 (OHB61-5 x OHB95-1A)	Urasaki et al. (2017)
11	1009	RAD	423 F_2 (K44 x Dali-11)	Cui et al. (2018)
20	2013	GBS	90 F_2 (DBGy-201 x Pusa Do Mousami)	Gangadhara Rao et al. (2018)

markers are applicable to genetic mapping of various traits. As mentioned above, it is still difficult to determine whether identified loci from independent mapping studies were identical or not, because of the difference in the mapping populations and correspondence of linkage groups and markers among different linkage maps. Nevertheless, it is important to know inheritance of various traits and the number of loci related to each phenotype, because genetic mapping of agronomic traits was applicable to marker-assisted breeding. Actually, several qualitative or quantitative traits were genetically mapped on each linkage map of bitter gourd.

In genetic mapping studies of bitter gourd, characteristics for flowers or fruits were focused, due to their agronomical importance. Fruit (epicarp) color decides its marketability; however, preference of color varies over regions. For example, green fruits are in demand in Southern China, whereas white fruits are preferred in Central China. Similarly, dark green to glossy green fruits are preferred in Northern India while white fruits are favored in Southern India (Srivastava and Nath 1972). Kole et al. (2012) found green fruit color is completely dominant over white color and spiny fruit surface structure to be completely dominant over smooth surface (Fig. 9.1). As a qualitative trait for sexual reproduction, gynoecy was supposed to be caused by any mutations in a single gene, although several gynoecious lines were independently reported in bitter gourd (Ram et al. 2006; Matsumura et al. 2014). Two independent study of gynoecy (Matsumura et al. 2014; Cui et al. 2018) showed that genetically mapped loci were located in the closed region at the end of linkage group. Another gynoecious locus was also mapped in the middle position of the linkage group, indicating that there are at least two loci for gynoecy in bitter gourd. Gynoecy in bitter gourd is essential for commercial F_1 breeding and gynoecious line with various phenotypes will be efficiently developed using markers linked to gynoecy (Table 9.2). In the fruits, color, luster, and surface structure were determined by one or two major loci, and their positions in the linkage map could be determined (Kole et al. 2012; Cui et al. 2018). These characteristics of fruits produce variation of commercial cultivars and markers for these loci will greatly contribute to breeding depending on consumer preference. Kole et al. (2012) also mapped qualitative loci controlling stigma color and seed color; however, stigma color was found to be monogenically controlled while seed color being oligogenically controlled (Fig. 9.1).

As quantitative traits, sex ratio (number of female flowers in a plant), first node of female flowers, days to first female flower, and several traits for fruits size were scored for mapping

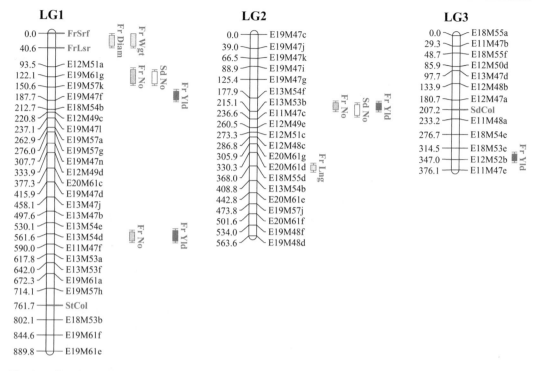

Fig. 9.1 Genetic linkage map of bitter gourd with positions of qualitative and quantitative trait loci. Qualitative traits are shown as markers in green text (Fr Srf, Fr Lsr, St Col, Sd Col and Fr Col designate fruit surface structure, fruit luster, stigma color, seed color and fruit color, respectively) and pattern in QTL bars are uniform for each quantitative trait (Fr Lng, Fr Diam, Fr Wgt, Fr No and Fr Yld designate fruit length, fruit diameter, fruit weight, fruit number, and fruit yield, respectively.) (Kole et al. 2012; reproduced with permission)

(Wang and Xiang 2013; Cui et al. 2018; Gangadhara Rao et al. 2018). These traits were directly effective to yield of bitter gourd fruits. Multiple QTLs for each trait were identified by independent studies, although some loci showing low LOD (<3.0) were included. By these studies, genetic relationships among traits were presumed, since QTLs for independently scored phenotypes were mapped at the close position, like fruit number per plant and fruit yield (Kole et al. 2012) or first female flower node and female flower number (Cui et al. 2018). It was still unknown how many QTLs for these traits were actually present in bitter gourd varieties, since correspondence of these independently identified QTLs were unknown. Because F_1 cultivars were commercially cultivated in bitter gourd, combining ability among inbred lines should be considered in its breeding program.

Thus, further detailed studies of these QTLs, like evaluation of multiple QTL combination, were essential for their practical use in marker-assisted breeding of bitter gourd.

As described above, molecular markers in the linkage map were assigned to genome sequence only in the two reports (Urasaki et al. 2017; Cui et al. 2018). From genetically mapped traits, causal genes can be determined by using corresponding genome sequence data. Particularly, qualitative traits, like fruit color or fruit surface structure, could be identified through corresponding genes by comparing their linked markers and reference genome sequences, because they were determined by major genes. For quantitative traits, allocation of QTL markers to reference genome will be helpful for understanding their genetic regulatory mechanism or networks.

Table 9.2 List of reported QTLs in bitter gourd

Reference	Traits	Number of QTL	Mapped linkage group[a]	LOD
Kole et al. (2012)	Fruit color	1	LG7	n.i.[b]
	Seed color	2	LG3	n.i.[b]
	Fruit surface structure	2	LG1	n.i.[b]
	Fruit length	2	LG2, LG7	3.02–3.53
	Fruit diameter	1	LG1	4.34
	Fruit weight	1	LG1	3.73
	Fruit number	4	LG1, LG2, LG5	6.47–7.52
	Fruit yield	4	LG1, LG2, LG3	3.32–8.12
Wang and Xiang (2013)	Female flower ratio	3	LG4, LG5, LG9	2.46–3.44
	First female flower node	3	LG4, LG5, LG9	2.47–4.71
	Fruit length	4	LG1, LG2, LG5, LG9	2.67–3.58
	Fruit diameter	5	LG1, LG9, LG11	3.65–4.78
	Fruit thickness	2	LG1	3.50–3.67
	Fruit shape	5	LG4, LG5, LG9, LG11	2.91–5.32
	Fruit pedicel length	3	LG4, LG8, LG9	2.47–4.20
	Fruit pedicel length ratio	5	LG4,LG6,LG8	2.53–4.85
	Fruit weight	4	LG4, LG5, LG6, LG12	2.75–3.27
	Fruit number per a plant	3	LG1, LG5	3.45–4.05
	Yield per a plant	2	LG5, LG9	2.38–3.43
	Stem diameter	2	LG2, LG4	6.57–7.07
	Internode length	2	LG2, LG5	2.54–4.31
Matsumura et al. (2014)	Gynoecy	1	LG1	n.i.[b]
Cui et al. (2018)	Gynoecy	2	MC01, MC02	4.43–6.46
	First female flower node	2	MC01	4.50–14.88
	Female flower number	2	MC01	7.99–25.12
	Fruit surface structure	2	MC04	5.39–37.07
	Immature fruit color	3	MC10	65.20–86.96
Gangadhara Rao et al. (2018)	Gynoecy	1	LG12	n.i.[b]
	First female flower node	5	LG5, LG9, LG14	2.9–4
	Days to first female flower	8	LG3,LG5, LG14, LG16	2.4–36.1
	Female flower ratio	9	LG9.LG13,LG14, LG16	2.5–6.4

[a]Name of linkage group was followed to the description in original publication
[b]n.i. means "not indicated"

9.5 Toward QTL Gene Cloning

The final goal of genetic mapping or QTL mapping is cloning of the causal gene(s) for the trait. The commonly employed strategy is map-based cloning method. In this strategy, target locus is narrowed down by fine-mapping and corresponding genomic region is defined. Finally, candidate gene and its sequence variant for the trait are identified. This analysis requires high-density maps and segregating individuals showing recombination at the target loci for narrowing down to the target region.

Recently, more rapid and direct strategy for causal gene identification of QTL was suggested as QTL-seq (Takagi et al. 2013). In QTL-seq, two groups of individuals (20-50) showing apparently opposite trait scores for each phenotype in F_2 or F_3 population were selected. Genomic DNA was pooled in each group from 20–50 individuals. Resequencing of these two bulked DNA was carried out and SNP-index (equivalent to allele frequency) was calculated at each SNP locus from read mapping data against reference genome. By plotting SNP-index at position of SNP locus in the reference genome, location of the gene for QTL could be visualized and the SNP in the causal gene of the target trait could be identified from gene model of the reference genome. In bitter gourd, once phenotypes of individuals were scored in segregated progeny, QTL-seq will be applicable, since reference genome sequence was already available. Particularly, qualitative traits, like fruit color or fruit surface characteristics, will be good targets for causal gene findings by QTL-seq method.

References

Baird NA, Etter PD, Atwood TS, Currey MC, Shiver AL, Lewis ZA, Selker EU, Cresko WA, Johnson EA (2008) Rapid SNP discovery and genetic mapping using sequenced RAD markers. PLoS ONE 3(10): e3376

Cui J, Luo S, Niu Y, Huang R, Wen Q, Su J, Miao N, He W, Dong Z, Cheng J, Hu K (2018) A RAD-based genetic map for anchoring scaffold sequences and identifying QTLs in bitter gourd (Momordica charantia). Front Plant Sci 9:477

Elshire RJ, Glaubitz JC, Sun Q, Poland JA, Kawamoto K, Buckler ES, Mitchell SE (2011) A robust, simple genotyping-by-sequencing (GBS) approach for high diversity species. PLoS ONE 6(5):e19379

Gangadhara Rao P, Behera TK, Gaikwad AB, Munshi AD, Jat GS, Boopalakrishnan G (2018) Mapping and QTL analysis of gynoecy and earliness in bitter gourd (Momordica charantia L.) using Genotyping-by-Sequencing (GBS) technology. Front Plant Sci 9:1555

Kole C, Gupta PK (2004) Genome mapping and map based cloning. In: Jain HK, Kharkwal MC (eds) Plant Breeding—From Mendelian to Molecular Approaches. Narosa Publishing House, New Delhi, India, pp 255–299

Kole C, Abbott AG (2008) Fundamentals of plant genome mapping. In: Kole C, Abbott AG (eds) Principles and practices of plant genomics. vol 1: Genome Mapping. Science Publishers—New Hampshire, Jersey, Plymouth, pp 1–22

Kole C, Olukolu BA, Kole P, Rao VK, Bajpai A, Backiyarani S, Singh J, Elanchezhian R, Abbott AG (2012) The first genetic map and positions of major fruit trait loci of bitter melon (Momordica charantia). J Plant Sci Mol Breed 1:1. https://doi.org/10.7243/2050-2389-1-1

Matsumura H, Miyagi N, Taniai N, Fukushima M, Tarora K, Shudo A, Urasaki N (2014) Mapping of the gynoecy in bitter gourd (Momordica charantia) using RAD-seq analysis. PLoS ONE 9(1):e87138

Ram D, Kumar S, Singh M, Rai M, Kalloo G (2006) Inheritance of gynoecism in bitter gourd (Momordica charantia L.). J Hered 97(3):294–295

Srivastava VK, Nath P (1972) Inheritance of some qualitative characters in Momordica charantia L. Indian J Hort 29:319–321

Takagi H, Abe A, Yoshida K, Kosugi S, Natsume S, Mitsuoka C, Uemura A, Utsushi H, Tamiru M, Takuno S, Innan H, Cano LM, Kamoun S, Terauchi R (2013) QTL-seq: rapid mapping of quantitative trait loci in rice by whole genome resequencing of DNA from two bulked populations. Plant J 74(1):174–183

Urasaki N, Takagi H, Natsume S, Uemura A, Taniai N, Miyagi N, Fukushima M, Suzuki S, Tarora K, Tamaki M, Sakamoto M, Terauchi R, Matsumura H (2017) Draft genome sequence of bitter gourd (Momordica charantia), a vegetable and medicinal plant in tropical and subtropical regions. DNA Res 24(1):51–58

Wang Z, Xiang C (2013) Genetic mapping of QTLs for horticulture traits in a $F_{2–3}$ population of bitter gourd (Momordica charantia L.). Euphytica 193(2):235–250

Genome Sequence of Bitter Gourd and Its Comparative Study with Other Cucurbitaceae Genomes

10

Hideo Matsumura and Naoya Urasaki

Abstract

Bitter gourd (*Momordica charantia*) is a diploid Cucurbitaceae species. It is grown in Asia, Africa and the Caribbean. Its genome information is expected to be quite important in elucidating its uniqueness and usefulness as a vegetable and medicinal plant. The first draft genome sequence of bitter gourd was determined as the reference genome and an improved version of its annotation is publicly available. Since it was presumed that divergence in bitter gourd varieties was relatively low, it might not be easy to identify molecular makers closely linked to a trait. By comparing genome and its annotation with those in other Cucurbitaceae species, bitter gourd was found to be phylogenetically distant from other known cucurbit crops and there are unique properties in encoding genes. Particularly, ribosome-inactivating protein (RIP) in bitter gourd was found to have antitumor or antiviral activities. Twice the number of RIP encoding genes was present in bitter gourd genome by comparing to other Cucurbitaceae genomes. These multiplicated and diverged RIP genes might characterize bitter gourd as the medicinal plant, while their biological functions were unknown.

10.1 Introduction

Bitter gourd (*Momordica charantia*) has unique characteristics in surface structure and bitter taste of the fruit, compared to other Cucurbitaceae crops, like melon, cucumber or squash. Additionally, it was known as a traditional medicinal plant species. In Cucurbitaceae, the draft genome sequence of cucumber was determined first (Huang et al. 2009). Subsequently, genome assemblies of melon and watermelon were reported (Garcia-Mas et al. 2012; Guo et al. 2013), and genes for unique and important traits in these crops have been identified. In 2017, the first draft genome sequence of bitter gourd was published (Urasaki et al. 2017) and their data is publicly available. Afterward, in accordance with the progress of technology and discount of the cost for next-generation sequencing (NGS), high-quality whole genome sequences were determined in several cucurbits (Montero-Pau et al. 2017; Sun et al. 2017; Wu et al. 2017).

In this chapter, information of bitter gourd draft genome and its annotation are presented. Also, similarities and uniqueness of bitter gourd genome or encoding genes are described by comparing with other Cucurbitaceae genomes.

H. Matsumura (✉)
Gene Research Center, Shinshu University, Ueda, Nagano, Japan
e-mail: hideoma@shinshu-u.ac.jp

N. Urasaki
Okinawa Agricultural Research Center, Itoman, Okinawa, Japan

C. Kole et al. (eds.), *The Bitter Gourd Genome*, Compendium of Plant Genomes,
https://doi.org/10.1007/978-3-030-15062-4_10

10.2 Sequencing and Assembly of Bitter Gourd Genome

As plant material for the first de novo sequencing analysis of bitter gourd genome, a monoecious inbred line, OHB3-1, was used (Urasaki et al. 2017). It was a line developed in Okinawa Agricultural Research Center, Japan, and its seeds are maintained by self-pollination for many years, expecting low heterozygosity. Whole genome sequencing by illumina sequencing is now the standard tool and its procedure has already been established. In the sequencing of OHB3-1 genome, several different types of libraries, including libraries of paired-end PCR-free and mate-pair with different sizes, were prepared. PCR-free library allows diminishing errors in the library preparation process and mate-pair libraries are essential for bridging assembled contigs, enhancing sizes of contigs or scaffolds. Usually, more than 100-fold coverage of sequence reads is necessary for de novo assembly of eukaryote genomes. Based on previously estimated genome size of bitter gourd (339 Mb; Urasaki et al. 2015), 37 Gb of sequence data was applied to de novo assembly in bitter gourd genome sequencing, which was equivalent to approximately 110 times that of the estimated genome size. In the recent NGS instruments, hundreds of giga base sequencing are not always difficult. On the other hand, choice of de novo assembly programs should be well considered in whole genome sequencing of higher plants or animals. Mostly, these programs are written using de Bruijn algorithm. In the OHB3-1 genome assembly, ALLPATH-LG assembler (Gnerre et al. 2011), which is expected to develop high-quality assembly, was employed. This assembler required sequence data for both fragment (paired-end) library and jumping (mate-pair) library.

The total length of the assembled scaffolds of OHB3-1 genome was 285.5 Mb, which comprised 1029 scaffolds, corresponding to approximately 84% of the previously estimated genome size (339 Mb). The N50 value of these scaffolds was 1.1 Mb, and the longest scaffold size was over 7 Mb. Compared to the previously predicted genome size of bitter gourd, about 15% of genome was not covered in this assembly. This was possibly due to redundant regions of the genome like repetitive sequences, and estimated genome size in the previous study might be ambiguous by the flow cytometric analysis. By Beschmarking Universal Single-Copy Orthologs (BUSCOs) assessment (Waterhouse et al. 2018) using Embryophyta odb9 dataset, 91.6% of core gene set was found in these scaffolds (Table 10.1), indicating that current assembly of OHB3-1 covered the whole genome at enough level.

De novo assembly of monoecious line 'Dali-11' genome was employed as the reference genome in RAD-seq read mapping (Cui et al. 2018), but its data are unpublished. In the public database, an additional genome assembly data of bitter gourd [Cultivar: TR(S108)] is registered (https://www.ncbi.nlm.nih.gov/assembly/GCA_900491585.1/). Although the detail of its analysis is still unknown, due to unpublished data, 296 Mb assembly, comprising of 3,101 scaffolds (N50 = 611 Kb) was obtained.

Table 10.1 Annotation summary of reference genome assembly (OHB3-1)

Number of *ab initio* predicted protein-coding genes[a]	45,859
Complete BUSCOs (percentage of coverage in total 1,440 BUSCO group)	1319 (91.6%)
Number of protein-coding genes[b]	19,431
(mean length of genes)	(4421 bp)
Number of CDSs[b]	28,666
(mean length of CDS)	(1412 bp)

[a]Urasaki et al. (2017)
[b]NCBI Momordica charantia Annotation Release 100

10.3   Annotation of Reference Genome Sequence

According to the gene prediction in OHB3-1 genome sequence by FGENESH program, 45,859 protein-coding genes were found (Table 10.1; Urasaki et al. 2017). Although predicted gene models from genome sequences are varied depending on employed programs and parameters for gene prediction, the number of ab initio predicted genes was usually overestimated due to probable pseudogenes or small open reading frames (ORFs).

Recently, more detailed annotation of OHB3-1 genome sequence assembly is released as NCBI Momordica charantia Annotation Release 100 (https://www.ncbi.nlm.nih.gov/genome/annotation_euk/Momordica_charantia/100/). Since models are determined by both Gnomon based on hidden Markov model (https://www.ncbi.nlm.nih.gov/genome/annotation_euk/gnomon/) and experimental data in this annotation, the number of genes or features was more accurate than that of genes only by ab initio prediction. In this annotation, 28,666 CDSs are count in the same reference genome and 19,431 protein-coding genes were defined (Table 10.1). These numbers and mean length of CDSs or genes are comparable to those in melon or cucumber. In this annotation, transcriptome (RNA-seq) data of bitter gourd tissues, registered in sequence read archive (SRA) database were also mapped in the draft genome sequence. Totally, 86% of RNA-seq reads in all the samples could be aligned. Except for 454 sequencing data of cDNA (Yang et al. 2010), 81–89% of RNA-seq reads were aligned to the genome, indicating that still around 15% of genome sequence might not be covered in the current assembly.

Of ab initio predicted 45,859 proteins, it was shown that 22,091 proteins harbored any known domains by InterProscan analysis (https://www.ebi.ac.uk/interpro/search/sequence-search). Average length of proteins with known domains was 1074 a.a., whereas proteins without domains showed much shorter length (233 a.a.), suggesting that actually expressed genes were less. Of all the identified domains (Table 10.2), reverse transcriptase (RNA-dependent DNA polymerase), probably encoded by retrotransposons, was the most frequent. Except for domains encoded by transposable elements, there are many proteins with pentatricopeptide repeat (PPR), protein kinase, P450, and zinc knuckle domain. Also, putative transcriptional factors are predominantly present, which encode several kinds of DNA-binding domains (Table 10.2). In spite of functional-domain-based categorizing of proteins in the recent annotation, the frequency of these major proteins might be conserved.

10.4   Sequence Divergence in Bitter Gourd Resources

Bitter gourd is mainly cultivated in south, south-east and east Asia. Many varieties are present in these regions, showing apparently different phenotypes like fruit shape or fruit color. Still, large-scale sequence divergence among these varieties was not well evaluated. By RAD-seq analysis, sequence divergence between two different inbred lines could be partially evaluated (Urasaki et al. 2017). In this study, approximately 3% of the uniquely reference-mapped RAD-tags in each inbred line contained mismatches and/or indels, suggesting an average frequency of these polymorphic loci as once per 2.7 kb in the bitter gourd genome. However, this evaluation did not always represent the actual frequency of polymorphisms, because all the RAD-tag sequences were not always covered in this analyzed data. Therefore, the genome of monoecious inbred line OHB95-1A was resequenced by paired-end reads of illumina HiSeq 2000 and reference mapping was carried out by Burrows-Wheeler Aligner (BWA). After single nucleotide polymorphism (SNP) and indel calling, 94,562 SNP and 41,295 indel loci were identified between OHB3-1 and OHB95-1A (Table 10.3). Focused on the homozygous polymorphisms, the density of SNP or indel was 3.6 kb or 7.7 kb, respectively. So far as the called indels by BWA, larger than 10 nt indels were limited to around 6% of detected indel loci (Table 10.3). It is difficult to compare

Table 10.2 The 20 most frequent observed domains in the predicted proteins in the bitter gourd reference genome

InterPro ID	Domain description	Count of proteins
IPR013103	Reverse transcriptase (RNA-dependent DNA polymerase)	1338
IPR000477	Reverse transcriptase (RNA-dependent DNA polymerase)	1087
IPR005162	Retrotransposon gag protein	924
IPR001584	Integrase core domain	916
IPR002885	PPR repeat	420
IPR000719	Protein kinase domain	366
IPR001128	Cytochrome P450	225
IPR026960	Zinc-binding in reverse transcriptase	206
IPR025724	GAG-pre-integrase domain	204
IPR001878	Zinc knuckle	192
IPR001245	Protein tyrosine kinase	188
IPR001005	Myb-like DNA-binding domain	179
IPR001611	Leucine-rich repeat	150
IPR000504	RNA recognition motif. (a.k.a. RRM, RBD, or RNP domain)	149
IPR001680	WD domain, G-beta repeat	142
IPR018289	MULE transposase domain	118
IPR011598	Helix-loop-helix DNA-binding domain	111
IPR001471	AP2 domain	110
IPR003653	Ulp1 protease family, C-terminal catalytic domain	108
IPR001841	Ring finger domain	108

Table 10.3 Summary of polymorphisms detected by read mapping of OHB95-1A genome against reference genome (OHB3-1)

Total number of sequence reads(M)[a]	406
SNP loci[b]	94,562
(homozygous)	78,661
(heterozygous)	15,901
Indel loci[b]	41,295
(homozygous)	36,874
(heterozygous)	4421
(>10 nt indels)	2371

[a]Paired-end sequence reads analyzed by Illumina HiSeq 2000
[b]SNPs and indels were called using bwa, samtools and bcftools

sequence divergency among other bitter gourd varieties. Genome-wide polymorphic loci, identified by RAD-seq or genotyping-by-sequencing (GBS) technology, were ranged between 1500 and 2000 loci (Cui et al. 2018; Gangadhara Rao et al. 2018), even when independent parental lines were compared. This suggested that nucleotide divergence might be almost similar in bitter gourd

resources. In cucumber, approximately 500 thousand SNPs and 50–60 thousand indels were detected in two inbred lines by reference mapping (Xu et al. 2016). Comparing to this, genetic divergence in bitter gourd seems to be low.

Low sequence divergence among bitter gourd inbred lines or cultivars restricts the application of molecular markers for selecting particularly traits in the breeding program. For an example, *GTFL-1* was found as a gynoecy-liked marker (Matsumura et al. 2014), and its genotype should be different between male and gynoecious female parent, for selection of gynoecious individuals using this marker. In fact, there are several monoecious domestic cultivars or inbred lines, showing identical genotype to gynoecious line (data not shown). Therefore, for applying molecular markers for breeding in wide varieties, causal genes for each qualitative or quantitative trait should be identified.

10.5 Comparative Study with Other Cucurbitaceae Genomes

In Cucurbitaceae, draft genome sequences of eight species (*Cucumis sativus. Cucumis melo, Citrulus lanatus, Cucurbita pepo, Cucurbita maxima, Cucurbita moschata, Lagenaria siceraria, and Momordica charantia*) were already published (Huang et al. 2009; Garcia-Mas et al. 2012; Guo et al. 2013; Montero-Pau et al. 2017; Sun et al. 2017; Urasaki et al. 2017; Wu et al.

2017) and their assembly data is publicly available (Table 10.4). The predicted genome size of these species is ranged from 283 to 454 Mb. In *C. pepo* and *L. siceraria*, assembly data covered >90% of predicted whole genomes. But, in other species, sequences of around 30–15% of genome were undetermined. As described above, this was possibly due to redundant region like repetitive sequences or transposon-rich regions, where accurate de novo assembly was interrupted. Differences of N50 in these assemblies were observed, although all assemblies showed >1 Mb of N50. Of these Cucurbitaceae genome assembly, bitter gourd reference genome still showed the smallest N50. Generally, N50 length depended on heterozygosity, insert size of sequence libraries, read length or parameters of assembly program. In de novo sequencing of bitter gourd, long-sized mate-pair libraries (15 or 20 kb) were not involved (Urasaki et al. 2017). Numbers of protein-coding genes of *C. melo, C. sativus,* and *M. charantia* (bitter gourd) were less than 20,000 genes and fewer than those in other cucurbits, although there were differences in the annotation procedures.

Table 10.4 Assembly summary of draft genome in Cucurbitaceae species

Species	Estimated genome size (Mb)	Total length of assembly (Mb)	N50 (kb)	Assembly level	Reference	Number of protein-coding genes
Cucumis melo	454	375	4,678	Chromosome	Garcia-Mas et al. (2012)	19,502[a]
Cucumis sativus	367	243.5	1,140	Chromosome	Huang et al. (2009)	18,738[a]
Citrullus lanatus	425	353.3	2,378	Chromosome	Guo et al. (2013)	23,440
Cucurbita pepo	283	263.5	1,750	Chromosome	Montero-Pau et al. (2017)	29,281[a]
Cucurbita maxima	380	271.4	3,717	Scaffold	Sun et al. (2017)	27,229[a]
Cucurbita moschata	372	269.9	3,996	Scaffold	Sun et al. (2017)	27,814[a]
Lagenaria siceraria	334	313.4	8,701	Scaffold	Wu et al. 2017	22,472
Momordica charantia	339	285.6	1,100	Scaffold	Urasaki et al. (2017)	19,431[a]

[a]The number of protein-coding genes was based on annotation report of genome assembly by NCBI

10.5.1 Syntey of Bitter Gourd and Other Species

Syntey among four Cucurbitaceae species including bitter gourd was analyzed using SyMap4.2 program (Urasaki et al. 2017). Synteny blocks, comprising at least seven anchored regions between two genome sequences, were seen in these Cucurbitaceae species, but their chromosome numbers are different and fragmented synteny blocks were mapped in each chromosome. Between bitter gourd and watermelon genome, synteny blocks of large size (>10 Mb) were found, compared between bitter gourd genome and melon or cucumber genomes, implying relative similarities between the bitter gourd and watermelon. In the synteny analysis, 14,775 gene loci presumed to be conserved in all compared Cucurbitaceae species. Of the genes at these loci, 69 loci were selected as unique conserved genes among all four species. Based on the alignment of encoded amino acid sequences of these genes, phylogenetic tree was constructed. Also, in this analysis, bitter gourd was relatively close to watermelon, rather than melon and cucumber, although it seemed to be evolutionary far from these species. In Cucurbitaceae species, which showed draft genome sequences, phylogenetic relationship was observed by aligning amino acid sequences chloroplastic matK (maturase K, Fig. 10.1). According to this tree, bitter gourd belongs to different group from Cucurbitaceae species, but was likely to be distant from other species.

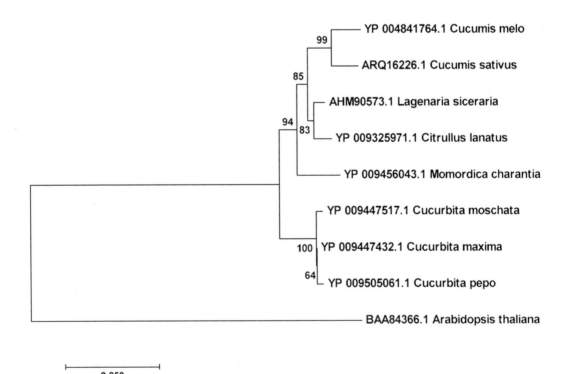

Fig. 10.1 Phylogenetic analysis of chloroplast matK (maturase K). gourd. Based on amino acid sequences of matK from eight Cucurbitaceae plants and Arabidopsis thaliana, phylogenetic tree was constructed using the Neighbor-Joining method by MEGA7.0.18. The percentage of replicate trees in which the associated taxa clustered together in the bootstrap test (1,000 replicates) is shown next to the branches. The tree is drawn to scale, with branch lengths in the same units as those of the evolutionary distances used to infer the phylogenetic tree. The evolutionary distances were computed using the p-distance method and are in the units of the number of amino acid differences per site

10.5.2 Diversity of *Momordica* Species

Schaefer and Renner (2010) studied molecular phylogenetic analysis of 122 accessions, including 47 *Momordica* species in Asia and Africa, based on organelle and nuclear genes. In these *Momordica* species, 23 monoecious species like bitter gourd and 24 dioecious species were included. They were separated in 11 clades in the tree and its origin was predicted to be Africa rather than Asia. Interestingly, in accordance with its dispersal from Africa to Asia, seven times conversion from dioecy to monoecy has occurred in this species. According to this result, comparative genomics study among monoecious and dioecious species in genus *Momordica* will elucidate the evolution of sex determination. In other *Momordica* species except for bitter gourd, genetic and genomic information were quite limited. According to the study of intraspecific crossability among several *Momordica* species (Bharathi et al. 2012), hybrid could be obtained from the cross between bitter gourd and monoecious *M. balsamina*, indicating their genome structure might be similar.

10.6 Orthologous Genes for Sex Determination in Cucurbitaceae

Cultivated Cucurbitaceae crops, including bitter gourd, mostly showed monoecious phenotype, which have male or female unisexual flowers in a plant. In melon and cucumber, their sex determination (male or female flower determination) mechanisms were extensively studied and these species are now model for elucidating monoecy in higher plants. For sex determination in these *Cucumis* species, it was demonstrated that ethylene is a key factor, and genes for its biosynthesis, aminocyclopropane-1-carboxylic acid (ACC) synthase genes, were identified as sex determination genes. *CmAcs11* was responsible for female flower determination in melon (Boualem et al. 2015), and *CmAcs-7*, as an additional ACC synthase gene, was also shown to regulate unisexual flower

development (Boualem et al. 2009). Their orthologous genes were also identified and functionally conserved in cucumber. Ethylene as a phytohormone was well known to have multiple functions in higher plants, including fruit ripening, senescence or biotic/abiotic stress response. In the plant genome, multiple copies of ACC synthase genes are present and each gene was supposed to have a different role. Phylogenetic analysis was carried out using putative orthologous proteins for CmAcs-7 and CmAcs11 in four Cucurbitaceae plants including bitter gourd (Urasaki et al. 2017). In this analysis, CmAcs-7 and its homologous proteins were separated from CmAcs11 and its homologs, predicting the functional differentiation of these two group of ACC synthases.

As a male determination gene, *CmWip1* was identified in melon (Martin et al. 2009). It encodes a Zn-finger DNA-binding protein and its loss of function caused gynoecy. Homologous gene to *CmWip1* was also found in bitter gourd and phylogenetic analysis of its encoding protein (XP_022148817.1 in current NCBI annotation) presumed to be an ortholog (Urasaki et al. 2017). However, its sequence alignment to *CmWip1* showed less similarity in the region out of conserved Zn-finger domain. This possible divergence of sex determination genes is described in sex determination chapter (see Chap. 6).

10.7 Genes for Functional Ingredients of Bitter Gourd

Bitter gourd was also studied as a medicinal plant, and its functional ingredients, particularly for antidiabetes, antitumor or antivirus, have been explored. Still, these ingredients are mostly organic compounds, and genes or proteins for their biosynthesis were almost unknown. However, a few kinds of proteins were shown to be functional ingredients. Trypsin inhibitors were suggested to have antidiabetic function (Lo et al. 2014) and ribosome-inactivating proteins showed antitumor and antiviral activities (Puri et al. 2009; Fan et al. 2015). According to the ab initio gene prediction, copies of genes encoding these proteins were more frequently present in bitter gourd genome than melon, cucumber, and watermelon genomes. It is possible that these genes characterize the uniqueness of bitter gourd as a medicinal plant.

In the recent annotation of the same draft genome sequence by NCBI, only two trypsin inhibitors were involved, and previously purified putative trypsin inhibitor proteins were not included. This is possibly due that their amino acid lengths were short (mostly, less than 100 a. a.) and they might be limitedly expressed in the specific tissues or stages.

10.7.1 Genes Encoding Ribosome-Inactivating Protein in Bitter Gourd Genome

Ribosome-inactivating protein (RIP) is a uniquely found protein in plants and two groups of RIPs are known; type 1 and type 2 (Walsh et al. 2013). In bitter gourd, some type 1 RIPs, as monomeric proteins, are suggested to have medicinal properties. Particularly, MAP30 and alpha-momorcharin are well studied as antitumor and antiviral RIPs in bitter gourd (Lee-Huang et al. 1995; Yao et al. 2011). Including MAP30, 17 type 1 ribosome-inactivating proteins (RIPs) were found in the NCBI Momordica charantia Annotation Release 100, although 18 genes encoding type 1 RIP were seen in the previous gene prediction by fgenesh. The presence of many copies of RIP genes was unique characteristics of bitter gourd genome and they were clustered in six scaffolds, which were nonsyntenic regions. In other Cucurbitaceae genomes, five to nine genes encoding putative RIPs were present (Table 10.5). Generally, RIPs in higher plants are presumed to represent a defensive function against pathogens or parasites (Barbieri et al. 2006; Puri et al. 2012). Still, it is unknown why significantly many RIP genes were present in bitter gourd. MAP30 and alpha-momorcharin were identical to beta-momorcharin and momordin I in the NCBI annotation, respectively. In alignment of amino acid sequence of these RIPs (Fig. 10.2), each protein showed sequence similarity to the other at less than 60% identities, while they carried RIP domain (Pfam00161). Also, it is undetermined whether all these RIPs have RNA N-glycosidase activity for cleaving ribosomal RNA. Even though sequences RIPs in bitter gourd were not always conserved well, amino acid sequence of MAP30 (beta-momorcharin) was identical to that of balsamin in *M. balsamina,* and trichosanthin (TCS) from *Trichosanthes kirilowii* was closely related to alpha-momorcharin (momordin I) in phylogenetic analysis of RIPs (Urasaki et al. 2017). Still, the required motif for antitumor or antiviral activities was not defined in these RIPs, but comparative studies of RIPs among bitter gourd varieties and related species will contribute to elucidation of their functions.

10.8 Future Prospects

Generally, the practical goal of whole genome sequence determination in eukaryote is chromosome-level genome assembly. For developing pseudomolecules as chromosome-level sequences, current version of bitter gourd draft genome sequence was still insufficient, its sequence did not probably cover whole genome and according to the correspondence to the linkage map of scaffolds via RAD-seq markers, there were some inconsistency in the marker order and unmapped scaffolds on the linkage map were observed (Urasaki et al. 2017). As the first approach, genome sequence assembly should be improved, resulting longer N50 and reduced total

Table 10.5 Number of genes encoding ribosome-inactivating protein in the genome of Cucurbitaceae species

C. melo[a]	C. sativus[a]	C. lanatus	C. pepo[a]	C. maxima[a]	C. moschata[a]	L. siceraria	M. charantia[a]
6	5	7	8	7	8	9	18

[a]Count from annotated proteins in annotation report on NCBI

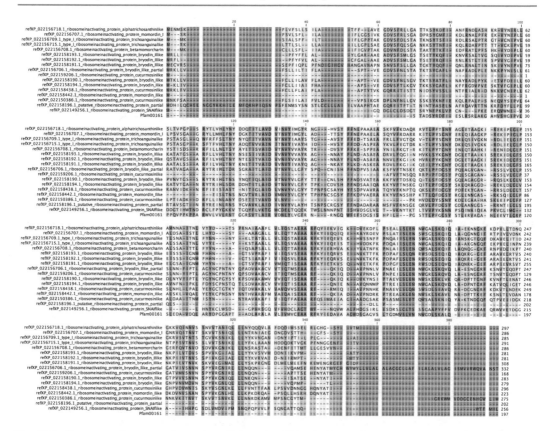

Fig. 10.2 Sequence alignment of putative ribosome-inactivating proteins is encoded in bitter gourd genome from NCBI Momordica charantia Annotation Release 100. Amino acid sequences of these proteins and Pfam00161 (ribosome-inactivating protein) domain were aligned by ClustalW

number of scaffolds or contigs. This challenge must be done by employment of long-read sequencing technologies, like PacBio or nanopore sequencing. In both technologies, >10 kb sequencing reads on average are practical and it was demonstrated that de novo assembly of eukaryote genome was dramatically improved. Particularly, low heterozygous genome of the inbred line will be effective material for obtaining high-quality genome assembly. Additional requirement is the high-density linkage map, where contigs were mapped and ordered in the linkage groups equivalent to chromosomes. Although there were two linkage maps, based on RAD-seq markers, consisted of 11 linkage groups (Urasaki et al. 2017; Cui et al. 2018), the number of mapped markers were not always distributed in all the scaffolds or contigs. Therefore, increasing of marker density is necessary for this purpose. Application of genome-wide SNP loci as markers should be considered to develop a hyperdense map using bitter gourd varieties with low sequence divergence.

Based on genome sequence information in bitter gourd, target genes for genetic improvement as a vegetable, and causal genes for functionality could be identified. In Cucurbitaceae crops, bitter gourd locates at a unique position phylogenetically, and its genome or gene structure will be helpful for elucidating its uniqueness and evolution in the future.

References

Barbieri L, Polito L, Bolognesi A, Ciani M, Pelosi E, Farini V, Jha AK, Sharma N, Vivanco JM, Chambery A, Parente A, Stirpe F (2006) Ribosome-inactivating proteins in edible plants and purification and characterization of a new ribosome-inactivating protein from *Cucurbita moschata*. Biochem Biophys Acta 1760(5):783–792

Bharathi LK, Munshi AD, Behera TK, Vinod, Joseph John K, Das AB, Bhat KV, Sidhu AS (2012) Production and preliminary characterization of inter-specific hybrids derived from *Momordica* species. Curr Sci 103(2):178–186

Boualem A, Troadec C, Camps C, Lemhemdi A, Morin H, Sari MA, Fraenkel-Zagouri R, Kovalski I, Dogimont C, Perl-Treves R, Bendahmane A (2015) A cucurbit androecy gene reveals how unisexual flowers develop and dioecy emerges. Science 350(6261): 688–691

Boualem A, Troadec C, Kovalski I, Sari MA, Perl-Treves R, Bendahmane A (2009) A conserved ethylene biosynthesis enzyme leads to andromonoecy in two *Cucumis* species. PLoS ONE 4(7):e6144

Cui J, Luo S, Niu Y, Huang R, Wen Q, Su J, Miao N, He W, Dong Z, Cheng J, Hu K (2018) A RAD-based genetic map for anchoring scaffold sequences and identifying QTLs in bitter gourd (*Momordica charantia*). Front Plant Sci 9:477

Fan X, He L, Meng Y, Li G, Li L, Meng Y (2015) Alpha-MMC and MAP30, two ribosome-inactivating proteins extracted from *Momordica charantia*, induce cell cycle arrest and apoptosis in A549 human lung carcinoma cells. Mol Med Rep 11(5):3553–3558

Gangadhara Rao P, Behera TK, Gaikwad AB, Munshi AD, Jat GS, Boopalakrishnan G (2018) Mapping and QTL analysis of gynoecy and earliness in bitter gourd (*Momordica charantia* L.) using Genotyping-by-Sequencing (GBS) technology. Front Plant Sci 9:1555

Garcia-Mas J, Benjak A, Sanseverino W, Bourgeois M, Mir G, González VM, Hénaff E, Câmara F, Cozzuto L, Lowy E, Alioto T, Capella-Gutiérrez S, Blanca J, Cañizares J, Ziarsolo P, Gonzalez-Ibeas D, Rodríguez-Moreno L, Droege M, Du L, Alvarez-Tejado M, Lorente-Galdos B, Melé M, Yang L, Weng Y, Navarro A, Marques-Bonet T, Aranda MA, Nuez F, Picó B, Gabaldón T, Roma G, Guigó R, Casacuberta JM, Arús P, Puigdomènech P (2012) The genome of melon (*Cucumis melo* L.). Proc Natl Acad Sci USA 109(29):11872–11877

Gnerre S, Maccallum I, Przybylski D, Ribeiro FJ, Burton JN, Walker BJ, Sharpe T, Hall G, Shea TP, Sykes S, Berlin AM, Aird D, Costello M, Daza R, Williams L, Nicol R, Gnirke A, Nusbaum C, Lander ES, Jaffe DB (2011) High-quality draft assemblies of mammalian genomes from massively parallel sequence data. Proc Natl Acad Sci USA 108 (4):1513–1518

Guo S, Zhang J, Sun H, Salse J, Lucas WJ, Zhang H, Zheng Y, Mao L, Ren Y, Wang Z, Min J, Guo X, Murat F, Ham BK, Zhang Z, Gao S, Huang M, Xu Y, Zhong S, Bombarely A, Mueller LA, Zhao H, He H, Zhang Y, Zhang Z, Huang S, Tan T, Pang E, Lin K, Hu Q, Kuang H, Ni P, Wang B, Liu J, Kou Q, Hou W, Zou X, Jiang J, Gong G, Klee K, Schoof H, Huang Y, Hu X, Dong S, Liang D, Wang J, Wu K, Xia Y, Zhao X, Zheng Z, Xing M, Liang X, Huang B, Lv T, Wang J, Yin Y, Yi H, Li R, Wu M, Levi A, Zhang X, Giovannoni JJ, Wang J, Li Y, Fei Z, Xu Y (2013) The draft genome of watermelon (*Citrullus lanatus*) and resequencing of 20 diverse accessions. Nat Genet 45 (1):51–58

Huang S, Li R, Zhang Z, Li L, Gu X, Fan W, Lucas WJ, Wang X, Xie B, Ni P, Ren Y, Zhu H, Li J, Lin K, Jin W, Fei Z, Li G, Staub J, Kilian A, van der Vossen EA, Wu Y, Guo J, He J, Jia Z, Ren Y, Tian G, Lu Y, Ruan J, Qian W, Wang M, Huang Q, Li B, Xuan Z, Cao J, Asan, Wu Z, Zhang J, Cai Q, Bai Y, Zhao B, Han Y, Li Y, Li X, Wang S, Shi Q, Liu S, Cho WK, Kim JY, Xu Y, Heller-Uszynska K, Miao H, Cheng Z, Zhang S, Wu J, Yang Y, Kang H, Li M, Liang H, Ren X, Shi Z, Wen M, Jian M, Yang H, Zhang G, Yang Z, Chen R, Liu S, Li J, Ma L, Liu H, Zhou Y, Zhao J, Fang X, Li G, Fang L, Li Y, Liu D, Zheng H, Zhang Y, Qin N, Li Z, Yang G, Yang S, Bolund L, Kristiansen K, Zheng H, Li S, Zhang X, Yang H, Wang J, Sun R, Zhang B, Jiang S, Wang J, Du Y, Li S (2009) The genome of the cucumber, *Cucumis sativus* L. Nat Genet 41(12):1275–1281

Lee-Huang S, Huang PL, Chen HC, Huang PL, Bourinbaiar A, Huang HI, Kung HF (1995) Anti-HIV and anti-tumor activities of recombinant MAP30 from bitter melon. Gene 161(2):151–156

Lo HY, Ho TY, Li CC, Chen JC, Liu JJ, Hsiang CY (2014) A novel insulin receptor-binding protein from *Momordica charantia* enhances glucose uptake and glucose clearance in vitro and in vivo through triggering insulin receptor signaling pathway. J Agri Food Chem 62(36):8952–8961

Martin A, Troadec C, Boualem A, Rajab M, Fernandez R, Morin H, Pitrat M, Dogimont C, Bendahmane A (2009) A transposon-induced epigenetic change leads to sex determination in melon. Nature 461 (7267):1135–1138

Matsumura H, Miyagi N, Taniai N, Fukushima M, Tarora K, Shudo A, Urasaki N (2014) Mapping of the gynoecy in bitter gourd (*Momordica charantia*) using RAD-seq analysis. PLoS ONE 9(1):e87138

Montero-Pau J, Blanca J, Esteras C, Martínez-Pérez EM, Gómez P, Monforte AJ, Cañizares J, Picó B (2017) An SNP-based saturated genetic map and QTL analysis of fruit-related traits in zucchini using genotyping-by-sequencing. BMC Genom 18(1):94

Puri M, Kaur I, Perugini MA, Gupta RC (2012) Ribosome-inactivating proteins: current status and biomedical applications. Drug Discov Today 17(13–14):774–783

Puri M, Kaur I, Kanwar RK, Gupta RC, Chauhan A, Kanwar JR (2009) Ribosome inactivating proteins (RIPs) from *Momordica charantia* for antiviral therapy. Curr Mol Med 9(9):1080–1094

Schaefer H, Renner SS (2010) A three-genome phylogeny of *Momordica* (Cucurbitaceae) suggests seven returns from dioecy to monoecy and recent long-distance dispersal to Asia. Mol Phylogenet Evol 54(2):553–560

Sun H, Wu S, Zhang G, Jiao C, Guo S, Ren Y, Zhang J, Zhang H, Gong G, Jia Z, Zhang F, Tian J, Lucas WJ, Doyle JJ, Li H, Fei Z, Xu Y (2017) Karyotype stability and unbiased fractionation in the paleo-allotetraploid cucurbita genomes. Mol Plant 10(10):1293–1306

Urasaki N, Takagi H, Natsume S, Uemura A, Taniai N, Miyagi N, Fukushima M, Suzuki S, Tarora K, Tamaki M, Sakamoto M, Terauchi R, Matsumura H (2017) Draft genome sequence of bitter gourd (*Momordica charantia*), a vegetable and medicinal plant in tropical and subtropical regions. DNA Res 24 (1):51–58

Urasaki N, Tarora K, Teruya K (2015) Comparison of genome size among seven crops cultivated in Okinawa. Bull Okinawa Pref Agri Res Ctr 9:47–50

Walsh MJ, Dodd JE, Hautbergue GM (2013) Ribosome-inactivating proteins. Virulence 4(8):774–784

Waterhouse RM, Seppey M, Simão FA, Manni M, Ioannidis P, Klioutchnikov G, Kriventseva EV, Zdobnov EM (2018) BUSCO applications from quality assessments to gene prediction and phylogenomics. Mol Biol Evol 35:543–548

Wu S, Shamimuzzaman M, Sun H, Salse J, Sui X, Wilder A, Wu Z, Levi A, Xu Y, Ling KS, Fei Z (2017) The bottle gourd genome provides insights into Cucurbitaceae evolution and facilitates mapping of a papaya ring-spot virus resistance locus. Plant J 92 (5):963–975

Xu Q, Shi Y, Yu T, Xu X, Yan Y, Qi X, Chen X (2016) Whole-genome resequencing of a cucumber chromosome segment substitution line and its recurrent parent to identify candidate genes governing powdery mildew resistance. PLoS ONE 11(10):e0164469

Yang P1, Li X, Shipp MJ, Shockey JM, Cahoon EB (2010) Mining the bitter melon (*Momordica charantia* L.) seed transcriptome by 454 analysis of non-normalized and normalized cDNA populations for conjugated fatty acid metabolism-related genes. BMC Plant Biol 10:250

Yao X, Li J, Deng N, Wang S, Meng Y, Shen F (2011) Immunoaffinity purification of a-momorcharin from bitter melon seeds (*Momordica charantia*). J Sep Sci 34(21):3092–3098

Toward Metabolomics in Bitter Gourd

11

Takeshi Furuhashi

Abstract

Bitter gourd, *Momordica charantia*, is often seen in the tropical dishes. It belongs to the family Cucurbitaceae. This plant has been investigated for unique bitter taste but also been utilized as a medicinal plant as folk medicine. In this chapter, previous studies on bitterness and on chemical studies for medicinal properties of bitter gourd are introduced. Bitter gourd secondary metabolites, such as flavonoids, phenolics, sterols, and terpenoids were already reported. Among these secondary metabolites, especially terpenoids are related to bitterness and pharmacology. It is reported that bitter gourd secondary metabolites can play an important role for plant interaction with pathogen. Indeed, some of them deter predators by using terpenoids. While its stem showed hypertrophy response in plant interaction with parasitic plant without showing any serious pathogenic response. As bitter gourd genome data are available now, further selection of target plants can be achieved. To select target plants, chemical assay is inevitable, and innovation of portable sensors will contribute to future application research.

11.1 Introduction

Bitter gourd, *Momordica charantia* L., belongs to the family Curcurbitaceae, which also includes cucumber (genus *Cucumis*), pumpkin (genus *Cucurbita*), watermelon (genus *Citrullus*), and calabash (genus *Lagenaria*). Its botanical name *Momordica* comes from the Latin word indicating "to bite" (Subratty et al. 2005). Many crops of the family Curcurbitaceae grow juicy fruits often utilized as food source. Bitter gourd is recognized as an important crop in tropical area (*Inter alia*, South Asian countries). Its fruits are typically used for dishes, as we see Goya-Chanpuru in Okinawa prefecture, Japan. There are some studies on food chemistry. Food chemical studies revealed that its fruits contain fatty acids (Yuwai et al. 1991), amino acids, and vitamins (especially vitamin C) (Goo et al. 2016). Many studies on bitter gourd are focused on bitterness-imparting chemical compounds and its application to medicine.

11.2 Bitterness in Bitter Gourd

Bitter taste of bitter gourd plant is conspicuous, while majority of other members in the family Cucubitaceae are not of bitter taste and even grow juicy sweet fruits. This calls for the pursuit of bitter taste. Indeed, there are some studies to investigate on the taste with different cultural conditions (Khandaker and Kotzen 2018).

T. Furuhashi (✉)
Anicom Specialty Medical Institute Inc, 8-17-1, Nishi Shinjuku, Shinjuku-ku, Tokyo, Japan
e-mail: takeshi.furuhashi@ani-com.com

However, one of the most important points is what are the main compounds underlying the bitter taste. These principles were already called as momordicine in an early study (Perry 1980). Since the 1980s, identification of these bitter principles and determination of their chemical structure have been performed (Yasuda et al. 1984). Based on alcohol solvent extraction and chemical structural elucidation by NMR, the main bitter principles were identified as triterpenoids. There are some candidates contributing bitterness of in bitter gourd which are, for instance, cucurbitane structure triterpenoids (e.g., momordicine, charantin) and phenolics (Haque et al. 2011; Kaur et al. 2013) (Fig. 11.1). Some of the terpenoids can be present as glucoside form.

In addition to the chemical approaches, an advance in genomics also enabled to investigate the genes related to control of bitterness and its evolutionary implication. One of the recent genomics studies on transcriptional factor (named *Bt*) clarified that bitterness in fruits has a plesiomorphic characteristics among Cucurbits,

and bitterness was reduced during the domestication process (Shang et al. 2014). In this study, cucurbitacin C was focused as the bitterness controlling compound in cucurbits. The oxidosqualene cyclase gene family encodes a cucurbitadienol synthase (named *Bi*) catalyzing the cyclization of 2,3-oxidosqualene which is the first critical path for cucurbitacin C biosynthesis. Genome-wide association study inferred that *Bt* activates *Bi* in fruits and thus controls bitterness.

It implies that people tried to develop non-bitter taste in crops as seen in cucumber, while bitter gourd still retains some pleiomorphic characteristics.

11.3 Bitter Gourd as Medicinal Plant

Not only important as a vegetable crop for food, bitter gourd is a unique plant with bitter taste that itself attracts many people. One of the main reasons that many people use bitter gourd as a

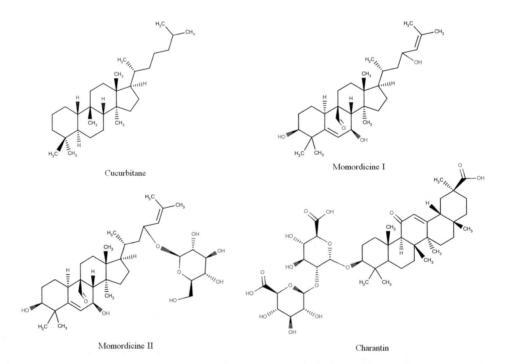

Fig. 11.1 Chemical structure of cucurbitane, momordicine I and II, charantin. Bitter principles identified from Cucubitaceae family possess cucurbitane structure-based chemicals. Some of them are triterpenoids with glucosides

food source is its potential herbal effect (Grover and Yadav 2004). A possibility of medicinal effect has been recognized since Ayurvedic age (Sarkar et al. 2015) rather than other traditional medicines (e.g., Chinese traditional medicine and *Unani* traditional medicine). Probably because one of the countries of origin for bitter gourd is India, it was developed as a traditional medicine in India at early stages being influenced by *Ayurveda*. In fact, even nowadays supplement using bitter gourd is commercially available, known as KANIM, which is based on *Ayurveda*.

In addition to India, bitter gourd plant has been used as a folk medicine in many countries, including Brazil, China, Cuba, Haiti, India, Mexico, Malaya, Nicaragua, Panama, Peru, and Trinidad (Kumar and Bhowmik 2010), as medicinal effects, antidiabetic, antihyperglycemic, antioxidant, antivirial, antimicrobial, anti-inflammatory, antifertility, and antitumor activities have been reported. On the other hand, an effect could be different based on conditions and it could be toxic or may have an adverse effect (Jia et al. 2017).

To date, attention and interest have been paid to actual chemicals contributing herbal or medical effect of bitter gourd plants. In the previous studies, it was reported that bitter gourd plants contain many secondary metabolites, such as phenolics, carotenoids, terpenoids, and sterols (Nagarani et al. 2014). *Inter alia*, fruits and juice are famous, but almost all parts of the plant (i.e., leaf, stem, root, fruits, and seeds) have been investigated and tested (Gupta et al. 2011). Based on bioactivity test, triterpenoids appear to contribute to impart medicinal effect, which are, for example, trinorcucurbitane and cucurbitane in bitter gourd root (Chen et al. 2008).

11.4 Host–Plant Interaction

Bitter gourd plants are sometimes recognized as agriculturally hardy plants. Agriculturally hardy often means that a plant can cope up with environmental stresses (i.e., abiotic stress) and can defend against pathogens (i.e., biotic stress) (Fig. 11.2). As bitter gourd contains many

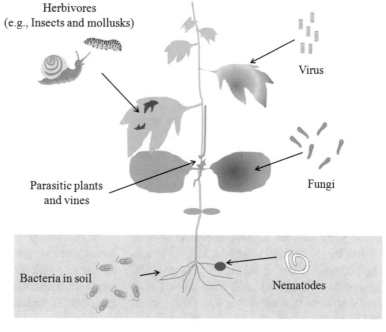

Fig. 11.2 Pathogens and foes for bitter gourd plants. Shoot part (especially leaves) of plants are attacked by herbivores (e.g., insects and snails), virus and fungi. Stem part are attacked by parasitic plant. Alternatively, plants are twisted by other vine trees and efficacy photosynthesis is reduced. In underground, there are many bacteria and nematodes, so that root would be damaged, or root-knot can be formed

Herbivores
(e.g., Insects and mollusks)

Virus

Parasitic plants
and vines

Fungi

Bacteria in soil

Nematodes

secondary metabolites, it is important to focus on biotic stress response and to investigate how these metabolites are used in host–plant interaction.

11.4.1 Pathogenic Response

Responding to lesions and pathogens is a feature common to all living organisms. Plants, as sessile organisms, have developed completely different defensive systems than those in animals (Lotze et al. 2007). For example, vertebrates have a cellular immune system based on T-cells and B-cells together with immunoglobulin-based antibody recognition. They have Toll-like receptor (TLR)-based PAMPs/MAMPs and DAMPs recognition, and the signals can be connected to adaptive responses. Plants, in contrast, lack such T-cells and have no antibody-based immune system or adaptive responses. Bacteria, fungi, herbivores, and viruses can be virulent and deleterious pathogens for plants (Baker et al. 1997; Howe and Jander 2008; Segonzac and Zipfel 2011). They can attack plant roots (e.g., cyst nematodes) and leaves (e.g., viruses). The relationship between plants and pathogens resembles a tug of war: plants have evolved special defensive systems, and pathogens then attempt to overcome them.

Enemies of bitter gourd plants certainly exist and plant–pathogen interaction has been reported. For example, infection of root-knot nematode (*Meloidogyne incognita*) causes serious damage to root (Singh et al. 2012). Bitter gourd appears to have evolved chemical defense rather than defense based on physically strong or hard organs. In fact, there is a report that some compounds can function as pathogenic response. α-momorcarin (α-MMC) is type I ribosome I-inactivating protein positively increased bitter gourd plant defense against cucumber mosaic viruses by enhancing jasmonic acid (JA) signaling pathway (Yang et al. 2011). This α-MMC present in bitter gourd seed also showed antifungal activity to *Fusarium solani* (Wang et al. 2016). In that paper, deformation of fungi cells with irregular budding and disruption of the fungal cell membrane were reported.

Secondary metabolites are important for plant interaction, such as allelochemicals. It seems that bitter gourd evolved secondary metabolite biosynthesis pathway to repel predators or to be distasteful for herbivores. Such chemicals for defense were found based on observation. Some beetles common in Cucurbitaceous crops (*Aulacophora femoralis, Epilachna admirabilis, and E. boisduvali*) seldom attack bitter gourd. From bitter gourd leaf methanol extract assay, momordicines were isolated, and a mixture of momordicine I and II were found to deter these insects (Abe and Matsuda 2000). A result indicates that bitter principle for human being can also be "bitter taste" for bitter gourd predators.

Moreover, given that most of the reported bitter gourd chemical defenses are shoot part (fruits and leaves), but chemical defense at root part might not be that strong. In fact, nemesis for bitter gourd appears to be rather present underground (such as root-knot nematodes).

11.4.2 Plant–Plant Interaction

Not only interaction with predators or foes, but there is interaction within plants. One is seen as techniques named "graft." Another one is interaction with parasitic plant interaction.

Graft is one of the interesting phenomena utilized in horticulture, although phenomenon itself can be seen in nature (Mudge et al. 2009). In general, graft is a combination of scion and rootstock, and plant which is strong to soil pathogens is selected as rootstock. By using graft technology, fruits quality, productivity, and disease resistance can be improved. Indeed, such techniques have improved quality of crops and foods, as we can see in citrus and apple tree (Goldschmidt 2013).

In the family Cucurbitaceae including bitter gourd, there are some researches on graft. One of the most problematic pests for bitter gourd is root-knot nematodes, thus selection of resistant rootstock is important. In a previous study, it turned out that one of the best rootstock candidates was pumpkin (*Cucurbita moschata*) in view of compatibility (Tamilselvi and Pugalendhi 2017).

Regarding metabolites' analysis, there is a *Cucurbita* vascular exudate metabolites profiling by GC-MS (Fiehn 2003). However, metabolomics approach has not been fully applied to bitter gourd graft issue.

Regarding plant pathogenic response, plant defense systems rely either on a performed defense (e.g., physical thickening of the cell wall) which is already developed before the infection, or on an inducible defense, which involves signal transduction leading to a hypersensitive response (HR) and systemic acquired resistance (SAR) (Greenberg 1997). In higher plants, a defense system eventually leads to programmed cell death (PCD), which can separate pathogen-infected or lesion areas from healthy, non-infected tissue. HR is non-autolytic rapid programmed cell death at an infected site, which prevents the spread of infection by pathogens in plants. The defensive system is a series of signal transductions from pathogen recognition to such HR and PCD (Berkey et al. 2012).

The trigger defense system functions mainly against microorganisms (e.g., bacteria and fungi), and little is known about host plants' special defense systems specifically targeted against parasitic plant. The holostemparasitic plant *Cuscuta* mainly parasitizes the stems of many different plants (Kuijt 1969). Because this parasite belongs to the plant kingdom, the defensive responses to it are probably not a simple analogy of the defensive system to other pathogens. In previous study, interaction between *Cuscuta* and bitter gourd was investigated, and a study uncovered that *Cucuta* parasitization caused hypertrophy of bitter gourd stem (hypocotyl part), as well as increase of vascular tissues (Fig. 11.3a and b) (Furuhashi et al. 2014). This phenomenon was implicated with increase of cytokinines and auxin. We found that *Cucumis* showed quite similar hypertrophy response against *Cuscuta* parasitization and thus hypertrophic response to parasitic plant *Cuscuta* would be characteristics in Cucurbitaceae. On the other hand, no serious defensive response to *Cuscuta*, such as HR was observed in all Cucurbitaceae species (e.g., *Cucumis* sp).

11.5 Conclusion and Future Perspective

In chemical studies on bitter gourd plant, one of the characteristics appears to be terpenoids biosynthesis pathway. On the other hand, it is common that terpenoids synthetic pathway is not shared with other primary metabolite pathways. As such, a cost of terpenoids' accumulation appears to be high for plant (Gershenzon 1994). Nonetheless, terpenoids often possess multifunction and could be important as hormone and pheromone effect. As an example, some of terpenoids synthetic pathways lead to plant hormone such as abscisic acid (ABA) (Cheng et al. 2007).

Another important aspect is that evolution of terpenoids biosynthesis is sometimes recognized as the outcome of plant–insect interaction. There are various types of terpenoids (monoterpene, diterpene, and triterpene) and some of them are volatiles. After terpenoids are synthesized in cytosol, terpenoids are accumulated in plastid and secreted (Trapp and Croteau 2001).

For instance, volatile organic compounds, *inter alia*, terpenoids can be used both to repel predators and to attract insects for pollination. Indeed, from Y-shaped glass tube olfactometer bioassay, it revealed that alkane present in bitter gourd flower surface wax attracted *Aulacophora foveicollis* (Mukherjee et al. 2013). It is interesting that bitter gourd possesses allelochemicals which can be used for pollination and attract insects, but at the same time, it possesses deterrent chemicals in leaves to repel some insects as herbivores.

In particular, bitter fruits taste based on terpenoids would be distasteful for some of the herbivores. Although these terpenoids are not lethal for many predators, it can be enough to increase the rate of survival for bitter gourd. Therefore, terpenoids' synthesis pathway would be evolutionary selected. In fact, implication between evolutionary plasticity in the plant kingdom and terpene synthase (TPS) has been discussed previously (Chen et al. 2011). Based on such information described above, how studies on bitter gourd will be in the future?

As bitter gourd plant genome data is now available, genomics approaches can be applied as

Fig. 11.3 Plant interaction between *Cuscuta* and bitter gourd. **a** The *Cuscuta* seedling before parasitization is 0.5 mm wide. Three days after parasitization, the bitter gourd stem started to swell (hypertrophy) and the *Cuscuta* seedling start elongating. One week later, the swollen bitter gourd stem doubled in width and the *Cuscuta* seedling reached about 20–30 cm in length. One month later, parasitized part was swollen more and even become huge cancer like bump with developing some *Cuscuta* shoot. Red arrow indicates bumps formed at bitter gourd stem.
b Transverse section of bitter gourd stem parasitized by *Cuscuta*. New vascular bundles are induced. At the initial stage, only eight vascular bundles are present. After 1 week, the number of vascular bundles increased near *Cuscuta* hyphae and the stem was swollen. Right bottom photograph also indicates increase of vascular bundles

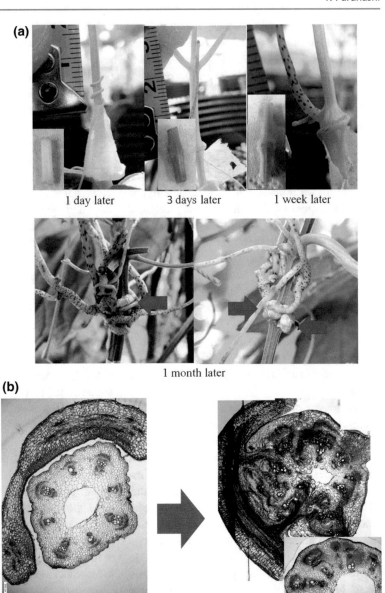

(a) 1 day later   3 days later   1 week later

1 month later

(b) 1 day later   1 week later

seen in other cucurbit genome-wide association studies (Shang et al. 2014). Application will mainly be bitterness control, pharmacology, and agriculture. For example, agriculturally hardy plants are important, meaning that bitter gourd which secretes secondary metabolites repelling predators without changing its taste would be desired.

Bitterness control also can be a tug of war, because one of the important aspects of bitter gourd is its herbal effect which tastes "bitter." This would be a plausible reason that bitter gourd has not been lost bitter taste and kept plesiomorphic characteristics during domestication, although other Cucurbitaceae crops lost bitterness and are being improved for their fruit quality. As such, it might be interesting to ask, can we develop "non-bitter bitter gourd"? There might be two choices. One is to reduce bitterness by innovating new cooking recipe. Nonetheless, reducing bitterness can be related to degradation of bioactive components. Another choice is genetically modified into non-bitter triterpenoids' synthesis without losing bioactivity. For achieving this, simple and easy triterpene assay kit is required. Current technical advance makes it possible to quantify triterpenes (e.g., charantin) from bitter gourd extract with simple colorimetric assay (Afthoni et al. 2018). Further advance is portable sensor techniques that would contribute to easier selection of desirable plants in the field.

References

Abe M, Matsuda K (2000) Feeding deterrents from *Momordica charantia* leaves to cucurbitaceous feeding beetle species. Appl Entomol Zool 35(1):143–149

Afthoni MH, Wahjudi M, Kuswandi B (2018) Novel scanometric assay for charantin in bitter melon (*Momordica charantia*) extract based on immobilized silver nitrate and methylene blue as colorimetric paper. J Food Chem Nanotechnol 4(2):27–32

Baker B, Zambryski P, Staskawicz B, Dinesh-Kumar SP (1997) Signaling in plant-microbe interactions. Science 276(2):726–733

Berkey R, Bendigeri D, Xiao S (2012) Sphingolipids and plant defense/disease: the "death" connection and beyond. Front Plant Sci 3(68):1–22

Cheng AX, Lou YG, Mao YB, Lu S, Wang LJ, Chen XY (2007) Plant terpenoids: biosynthesis and ecological functions. J Integr Plant Biol 49(2):179–186

Chen J, Tian R, Qiu M, Lu L, Zheng Y, Zhang Z (2008) Trinorcucurbitane and cucurbitane triterpenoids from the roots of *Momordica charantia*. Phytochemistry 69:1043–1048

Chen F, Tholl D, Bohlmann J, Pichersky E (2011) The family of terpene synthases in plants: a mid-size family of genes for specialized metabolism that is highly diversified throughout the kingdom. Plant J 66:212–229

Fiehn O (2003) Metabolic networks of *Cucurbita maxima* phloem. Phytochemistry 62:875–886

Furuhashi T, Kojima M, Sakakibara H, Fukushima A, Hirai MY, Furuhashi K (2014) Morphological and plant hormonal changes during parasitization by *Cuscuta japonica* on *Momordica charantia*. J Plant Interact 9(1):220–232

Gershenzon J (1994) Metabolic costs of terpenoid accumulation in higher plants. J Chem Ecol 20(6):1281–1328

Goldschmidt EE (2013) The evolution of fruit tree productivity: a review. Econ Bot 67(1):51–62

Goo KS, Ashari S, Basuki N, Sugiharto AN (2016) The bitter gourd *Momordica charantia* L.: morphological aspects, charantin and vitamin C contents. J Agri Vet Sci 9(10):76–81

Greenberg JT (1997) Programmed cell death in plant-pathogen interactions Annu Rev Plant Physiol Plant Mol Biol 48:525–545

Grover JK, Yadav SP (2004) Pharmacological actions and potential uses of *Momordica charantia*: a review. J Ethnopharmacol 93:123–132

Gupta M, Sharma S, Gautam AK, Bhadauria R (2011) *Momordica charantia* Linn. (Karela): nature" silent healer. Int J Pharmaceut Sci Rev Res 11(1):32–37

Haque ME, Alam MB, Hossain MS (2011) The efficacy of Cucurbitane type triterpenoids, glycosides and phenolic compounds isolated from *Momordica chranatia*: a review. Int J Pharmaceu Sci Res 2(5):1135–1146

Howe GA, Jander G (2008) Plant immunity to insect herbivores. Annu Rev Plant Biol 59:41–66

Jia S, Shen M, Zhang F, Xie J (2017) Recent advances in *Momordica charantia*: functional components and biological activities. Int J Mol Sci 18(2555):1–25

Kaur M, Deep G, Jain AK, Raina K, Agarwal C, Wempe MF, Agarwal R (2013) Bitter melon juice activates cellular energy sensor AMP-activated protein kinase causing apoptotic death of human pancreatic carcinoma cells. Carcinogenesis 34(7):1585–1592

Khandaker M, Kotzen B (2018) Taste testing bitter gourd (*Momordica charantia*) grown in aquaponics. Ecocycles 4(2):19–22

Kuijt J (1969) The biology of parasitic flowering plants. University of California Press, Berkley

Kumar SKP, Bhowmik D (2010) Traditional medicinal uses and therapeutic benefits of *Momordica charantia* Linn. Int J Pharmaceut Sci Rev Res 4(3):23–28

Lotze MT, Zeh HJ, Rubartelli A, Sparvero LJ, Amoscato AA, Washburn NR, DeVera ME, Liang X, Tör M, Billiar T (2007) The grateful dead: damage-associated molecular pattern molecules and reduction/oxidation regulate immunity. Immunol Rev 220:60–81

Mudge K, Jcinick J, Scofield S, Goldschinidt EE (2009) History of grafting. In: Janick J (ed) Horticultural reviews, vol 35. Wiley, pp 437–493

Mukherjee A, Sarkar N, Barik A (2013) Alkanes in flower surface waxes of *Momordica cochinchinensis* influence attraction to *Aulacophora foveicollis* Lucas (Coleoptera: Chrysomelidae). Neotrop Entomol 42:366–371

Nagarani G, Abirami A, Siddhuraju P (2014) Food prospects and nutraceutical attributes of *Momordica* species: a potential tropical bioresources—a review. Food Sci Hum Wellness 3:117–126

Perry LM (1980) Medicinal plants of east and southeast Asia, attributed properties and uses. MIT Press, Cambridge, p 117

Sarkar P, Kumar DHL, Dhumal C, Panigrahi SS, Choudhary R (2015) Traditional and ayurvedic foods of Indian origin. J Ethnol Foods 2:97–109

Segonzac C, Zipfel C (2011) Activation of plant pattern-recognition receptors by bacteria. Curr Opin Microbiol 14:54–61

Shang Y, Ma Y, Zhou Y, Zhang H, Duan L, Chen H, Zeng J, Zhou Q, Wang S, Gu W, Liu M, Ren J, Gu X, Zhang S, Wang Y, Yasukawa K, Bouwmeester HJ, Qi X, Zhang Z, Lucas WJ, Huang S (2014) Biosynthesis, regulation, and domestication of bitterness in cucumber. Science 346:1084–1088

Singh SK, Conde B, Hodda M (2012) Root-knot nematode (*Meloidogyne incognita*) on bitter melon (*Momordica charantia*) near Darwin, Australia. Australas Plant Dis 7:75–78

Subratty AH, Gurib-Fakim A, Mahomoodally F (2005) Bitter melon: an exotic vegetable with medicinal values. Nutr Food Sci 35:143–147

Tamilselvi NA, Pugalendhi L (2017) Graft compatibility and anatomical studies of bitter gourd (*Momordica charantia* L.) scions with Cucurbitaceous rootstocks. Int J Curr Microbiol Appl Sci 6(2):1801–1810

Trapp SC, Croteau RB (2001) Genomic organization of plant terpene synthases and molecular evolutionary implications. Genetics 158:811–832

Wang S, Zheng Y, Xiang F, Li S, Yang G (2016) Antifungal activity of *Momordica charantia* seed extracts toward the pathogenic fungus *Fusarium solani* L. J Food Drug Anal 24:881–887

Yang DH, Hettenhausen C, Baldwin IT, Wu J (2011) The multifaceted function of BAK1/SERK3. Plant immunity to pathogens and responses to insect herbivores. Plant Sig Behav 6(9):1322–1324

Yasuda M, Iwamoto M, Okabe H, Yamauchi T (1984) Structures of momordicines I, II and III. The bitter principles in the leaves and vines of *Momordica charantia* L. Chem Pharm Bull 32(5):2044–2047

Yuwai KE, Rao S, Kaluwin JC, Jones GP, Rivetts DE (1991) Chemical composition of *Momordica charantia* L. fruits. J Agri Food Chem 39:1762–1763

Future Prospects of Genomics and Breeding in Bitter Gourd

12

Hideo Matsumura, Tusar Kanti Behera
and Chittaranjan Kole

Abstract

Draft genome sequence of bitter gourd was currently released, but further improvements of assembly are required in its total length and number of scaffolds. Recent advances in the long-read sequencing technologies will allow to obtain the whole genome sequence of bitter gourd at chromosome level. Complete genome sequence and high-density linkage map are greatly helpful for finding genes for agronomic traits or biosynthesis of medicinal substances. Once candidate genes are found by genomic and genetic analyses, their biological functions should be elucidated. For this purpose, establishment of more efficient and easier method of genetic transformation is the future issue in bitter gourd. Genome information in bitter gourd must contribute to the study of intra- and inter-species genetic diversity in the *Momordica* species.

H. Matsumura
Gene Research Center, Shinshu University, Ueda, Nagano, Japan

T. K. Behera
Division of Vegetable Science, ICAR-Indian Agricultural Research Institute, Pusa, New Delhi 110012, India

C. Kole (✉)
ICAR-National Institute for Plant Biotechnology, Pusa, New Delhi 110012, India
e-mail: ckoleorg@gmail.com

It is expected that the draft genome sequence of bitter gourd would promote gene findings, DNA marker exploration, and extensive study of genetic diversity, although available genomic and genetic information in bitter gourd are still insufficient, compared with melon or cucumber.

Currently, the released version of draft genome sequence in bitter gourd covers almost the whole genome region and its annotation is also available (Urasaki et al. 2017). Actually, its total length (286 Mbp) was smaller than its expected genome size (>300 Mbp) and constituted from more than 1000 scaffolds. Therefore, further sequencing effort is needed for accomplishing complete genome sequence of the whole 11 chromosomes. Recently, whole genome sequencing at chromosomal level is feasible in various life organisms, even with large genome size. For these studies, long-read (>10 kb) sequencing like PacBio or nanopore sequencing technology is essential, which allows to develop large contigs by de novo assembling. In addition to improvement of assembly by long sequence reads, several strategies for bridging contigs are also employed for finalizing genome sequence. Optical mapping is a technology for constructing the physical map of whole genome, which is helpful for developing pseudomolecules. Actually, improved melon genome sequence could be obtained by mapping previously assembled scaffolds onto the optical map (Ruggieri et al. 2018). Hiâ€"C is a method for analyzing the proximity of genomic regions by sequencing of fragmented crosslinked DNA, which allows knowing spatial arrangements among contigs (Dudchenko et al. 2017). Because no additional instruments were needed for Hi‐C analysis, its application as a scaffolding tool of contigs is increasing. Apart from these methods by analyzing physical structure of the genome, the linkage map based on genetic recombination is also helpful for aligning contigs. Since two high-density linkage maps were already obtained based on RAD-seq or genotyping by sequencing (GBS) markers in bitter gourd (Urasaki et al. 2017; Cui et al. 2018), by allocating larger contigs derived from assembly of long sequence reads to these maps, pseudomolecules will be developed.

Whole genome sequence at chromosome level or high-density linkage map would promote genetic mapping of various characteristics in bitter gourd, including agronomic traits. In those traits, their causal genes or tightly linked DNA markers will be explored, which contribute to the marker-assisted breeding. Since it is difficult to identify causal genes for unique characteristics or traits in bitter gourd from homologs in other plant species, genetic mapping strategy must be greatly helpful. Currently, genetically mapped traits or quantitative trait loci (QTLs) in bitter gourd are mainly concerned to fruits, yield, or sex determination (Kole et al. 2012; Wang and Xiang 2013; Matsumura et al. 2014; Cui et al. 2018; Gangadhara Rao et al. 2018). Production of unique metabolites or functional ingredients characterizes bitter gourd as the medicinal plant. Biochemical approaches were mainly employed for elucidating its physiological functions, while its genetic or molecular genetic studies were limited. Therefore, mapping of genes for production of functional metabolites or proteins will definitely contribute to dissect their biosynthesis pathways or to find their regulatory genes. As a vegetable crop, resistance to biotic or abiotic stress is also quite important in bitter gourd cultivation, and QTL mapping of these traits will promote breeding various resistant cultivars.

Increasing adequacy of genome information in bitter gourd, causal or regulatory genes for various traits will be found as described above. However, there is still an obstacle to analyze biological functions of those identified genes in the bitter gourd plant. A few protocols for genetic transformation of bitter gourd were reported (Sikdar et al. 2005; Thiruvengadam et al. 2013; Narra et al. 2018), but efficiency of developing transgenic plants is still low in Cucurbitaceae species. In melon or cucumber, genetic mutants having independent alleles of the target genes were collected and their phenotypes were evaluated for proving their biological function in the plants (Martin et al. 2009; Boualem et al. 2014), instead of transgenic analysis. However, employment of similar strategy needs a lot of effort and time for growing and screening a large number of randomly mutagenized plants. Once more efficient and easier protocols for genetic transformation in bitter gourd are established, CRISPR/Cas-based technologies are now available to make the targeted loss of function or single-base editing of the genes efficiently, as shown in other plant species. Improvement of efficient methods for functional analysis of genes is necessary for further study in bitter gourd.

Bitter gourd is mainly cultivated in countries in south, south-east, and east Asia. In current studies, although DNA polymorphisms were explored as DNA markers in several bitter gourd cultivars or varieties (Kole et al. 2012; Wang and Xiang 2013; Matsumura et al. 2014; Cui et al. 2018; Gangadhara Rao et al. 2018), genetic diversity was still unknown at species level. Since reference genome is greatly helpful for elucidating sequencing-based genetic diversity. Not only for elucidating diversification or evolution of bitter gourd, those studies will contribute to choose appropriate parental lines for F_1 breeding program, which cause heterosis. Furthermore, accumulation of single nucleotide polymorphisms (SNPs) data may facilitate genetic association mapping of particular traits in bitter gourd resources in the future.

In *Momordica* species, cultivated species other than bitter gourd (*M. charantia*) are limited, but they may have potentials as medicinal or edible plants. Also, since monoecy and dioecy were diverged in *Momordica* species (Schaefer and Renner 2010), analysis of their causal genes will reveal evolution of sex determination in this genus. Thus, accelerated genome sequencing technology allows illuminating basic or applied science in *Momordica* species.

References

Boualem A, Fleurier S, Troadec C, Audigier P, Kumar AP, Chatterjee M, Alsadon AA, Sadder MT, Wahb-Allah MA, Al-Doss AA, Bendahmane A (2014) Development of a *Cucumis sativus* TILLinG platform for forward and reverse genetics. PLoS ONE 9:e97963

Cui J, Luo S, Niu Y, Huang R, Wen Q, Su J, Miao N, He W, Dong Z, Cheng J, Hu K (2018) A RAD-based genetic map for anchoring scaffold sequences and identifying QTLs in bitter gourd (*Momordica charantia*). Front Plant Sci 9:477

Dudchenko O, Batra SS, Omer AD, Nyquist SK, Hoeger M, Durand NC, Shamim MS, Machol I, Lander ES, Aiden AP, Aiden EL (2017) De novo assembly of the Aedes Aegypti genome using Hiâ€"C yields chromosome-length scaffolds. Science 356: 92–95

Gangadhara Rao P, Behera TK, Gaikwad AB, Munshi AD, Jat GS, Boopalakrishnan G (2018) Mapping and QTL analysis of gynoecy and earliness in bitter gourd (*Momordica charantia* L.) using genotyping-by-sequencing (GBS) technology. Front Plant Sci 9:1555

Kole C, Olukolu BA, Kole P, Rao VK, Bajpai A, Backiyarani S, Singh J, Elanchezhian R, Abbott AG (2012) The first genetic map and positions of major fruit trait loci of bitter melon (*Momordica charantia*). J Plant Sci Mol Breed 1:1. https://doi.org/10.7243/2050-2389-1-1

Martin A, Troadec C, Boualem A, Rajab M, Fernandez R, Morin H, Pitrat M, Dogimont C, Bendahmane A (2009) A transposon-induced epigenetic change leads to sex determination in melon. Nature 461:1135–1138

Matsumura H, Miyagi N, Taniai N, Fukushima M, Tarora K, Shudo A, Urasaki N (2014) Mapping of the gynoecy in bitter gourd (*Momordica charantia*) using RAD-seq analysis. PLoS ONE 9(1):e87138

Narra M, Ellendula R, Kota S, Kalva B, Velivela Y, Abbagani S (2018) Efficient genetic transformation of *Momordica charantia* L. by microprojectile bombardment. Biotechnol 8:2

Ruggieri V, Alexiou KG, Morata J, Argyris J, Pujol M, Yano R, Nonaka S, Ezura H, Latrasse D, Boualem A, Benhamed M, Bendahmane A, Cigliano RA, Sanseverino W, PuigdomÃ¨nech P, Casacuberta JM, Garcia-Mas J (2018) An improved assembly and annotation of the melon (*Cucumis melo* L.) reference genome. Sci Rep 8:8088

Schaefer H, Renner SS (2010) A three-genome phylogeny of *Momordica* (Cucurbitaceae) suggests seven returns from dioecy to monoecy and recent long-distance dispersal to Asia. Mol Phylogenet Evol 54(2):553–560

Sikdar B, Shafiullah M, Chowdhury AR, Sharmin N, Nahar S, Joarder OI (2005) Agrobacterium-mediated GUS expression in bitter gourd (*Momordica charantia* L.). Biotechnology 4:149–152

Thiruvengadam M, Praveen N, Chung IM (2013) An efficient Agrobacterium tumefaciens -mediated genetic transformation of bitter melon (*Momordica charantia* L.). Aust J Crop Sci 6:1094–1100

Urasaki N, Takagi H, Natsume S, Uemura A, Taniai N, Miyagi N, Fukushima M, Suzuki S, Tarora K, Tamaki M, Sakamoto M, Terauchi R, Matsumura H (2017) Draft genome sequence of bitter gourd (Momordica charantia), a vegetable and medicinal plant in tropical and subtropical regions. DNA Res 24 (1):51–58

Wang Z, Xiang C (2013) Genetic mapping of QTLs for horticulture traits in a F_{2-3} population of bitter gourd (*Momordica charantia* L). Euphytica 193(2):235–250

Printed in the United States
by Baker & Taylor Publisher Services